S0-BZW-982

BOOK SALE

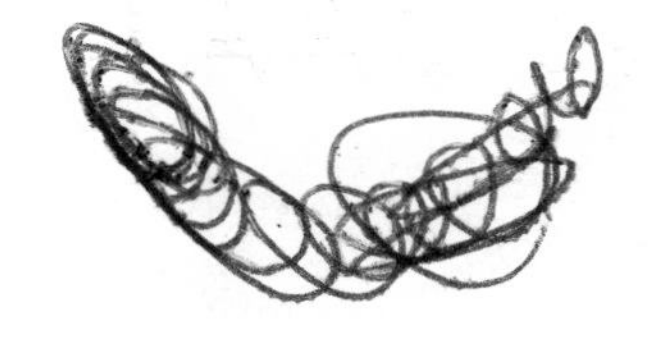

CAMBRIDGE STUDIES
IN MATHEMATICAL BIOLOGY: 2

Editors
C. CANNINGS
Department of Probability and Statistics, University of Sheffield
F. HOPPENSTEADT
Department of Mathematics, University of Utah

MECHANICS OF SWIMMING AND FLYING

STEPHEN CHILDRESS
New York University, Courant Institute of Mathematical Sciences

Mechanics of swimming and flying

CAMBRIDGE UNIVERSITY PRESS
Cambridge
London New York New Rochelle
Melbourne Sydney

Published by the Press Syndicate of the University of Cambridge
The Pitt Building, Trumpington Street, Cambridge CB2 IRP
32 East 57th Street, New York, NY 10022, USA
296 Beaconsfield Parade, Middle Park, Melbourne 3206, Australia

© Cambridge University Press 1981

First published 1981

Printed in the United States of America

Typeset by University Graphics, Inc., Atlantic Highlands, N.J.
Printed and bound by Vail-Ballou Press, Inc., Binghamton, N.Y.

Library of Congress Cataloging in Publication Data

Childress, Stephen.

Mechanics of swimming and flying.

(Cambridge studies in mathematical biology)

Bibliography: p.

Includes index.

1. Animal flight–Mathematical models.
2. Animals swimming–Mathematical models.
3. Fluid dynamics–Mathematical models.
4. Navier-Stokes equations.

I. Title. II. Series.
QP310.F5C48 1981 591.1′852 80–23364
ISBN 0 521 23613 4 hard covers
ISBN 0 521 28071 0 paperback

TO MY PARENTS
DORIS AND VIRGIL CHILDRESS

CONTENTS

PREFACE

This monograph developed from a course given at the Courant Institute in the spring of 1976. At that time a preliminary set of notes was prepared for the class by M. Lewandowsky, who also contributed Chapter 4 and made numerous helpful suggestions. My purpose at the time was to provide a brief complement to the relevant portions of Lighthill's book, *Mathematical Biofluiddynamics,* as well as a guide to the papers contained in the symposium proceedings *Swimming and Flying in Nature,* edited by Wu, et al, and to incorporate background and supplementary material as needed. The present book has been changed only by the addition of Chapter 12 on interactions and a number of revisions and corrections.

Although, as a result of these origins, this monograph falls far short of a definitive treatise, I hope that this glimpse of a fascinating and rapidly evolving body of research will be useful to students of fluid mechanics seeking a compact introduction to biological modeling. Those familiar with the aerodynamics of fixed-wing aircraft or the hydrodynamics of ships, to take two examples from many, are well aware that in these applications the theory is concerned with design as well as analysis. The products of our technology are both the subjects and the results of our mathematical modeling. In "natural" swimming and flying the situation is no longer the same. Nature has already provided the answers. We see around us the products of evolution and our task is to understand these "solutions" in the context of the presumably universal laws of mechanics. Some of our old ideas may be very useful, and it is precisely in those cases that our conclusions will probably be most secure. But the excitement lies near the boundaries of our perception, with the very small, the very fast, the elusive creatures of air and sea whose movements respect our mechanics but not necessarily our theories.

It is, therefore, not unexpected that analysis of natural swimming and flying is, from the standpoint of classical fluid mechanics, a difficult subject. Typically the geometry is complicated, the flow field highly nonstationary, the Reynolds number range awkward. But then our ultimate aim is understanding rather than design, and in many cases precise modeling is less useful than qualitative analysis grounded on very basic fluid mechanics. I have therefore attempted to give a brief introduction to concepts of general applicability as well as to go through, where important, more elaborate calculations in specific models. The material should be accessible to upper-level undergraduates and first-year graduate students with a firm understanding of introductory fluid mechanics. I hope that the monograph will also prove useful to others in its original context, as a supplementary reference in a special topics course.

If these aims are even approximately met, credit is due in no small part to the expert assistance I have had with the manuscript. F. C. Hoppensteadt made a careful reading and suggested a number of improvements; M. Croner typed the working draft; and the talented staff at the New York office of Cambridge University Press have made the final editing almost pleasurable.

Finally I would like to acknowledge with gratitude the support of the National Science Foundation under grant NSF-MCS75-09837-a01, and of the Guggenheim Foundation.

Stephen Childress

New York City
January, 1981

1

General considerations

1.1 Introduction

We shall deal with the fluid dynamics of the swimming of microorganisms and fish, and of the flying of birds and insects. In spite of the enormous range of the subject, there is a unifying theoretical framework in the nature of the underlying fluid-dynamical problem. The models we study are based on a description of the medium in which the organism moves. The motion of the medium is determined by the Navier–Stokes equations in the form

$$\frac{\partial \mathbf{u}}{\partial t} + \mathbf{u} \cdot \nabla \mathbf{u} + \frac{1}{\rho} \nabla p - \nu \nabla^2 \mathbf{u} = \mathbf{g} \tag{1.1a}$$

$$\nabla \cdot \mathbf{u} = 0 \tag{1.1b}$$

which are to be satisfied in the region exterior to the organism. A fluid described by (1.1 a) and (1.1 b) has a constant density ρ and a constant kinematic viscosity ν. Its state at any given instant is described by the pressure field $p(\mathbf{r}, t)$ and by the velocity field $\mathbf{u} = (u(\mathbf{r}, t), v(\mathbf{r}, t), w(\mathbf{r}, t))$, where $\mathbf{r} = (x, y, z) = (x_i, i = 1, 2, 3)$ is a point of space and t is time. In connection with problems of flight and the swimming of some fish, where the organisms are not neutrally buoyant, the action of a uniform gravitational field (with acceleration $\mathbf{g}$) must be recognized and is included on the right of (1.1 a). For the most part, however, gravity will not play an essential role in our work.

An "organism" can, for the most part, be taken to be an impenetrable region (generally of constant volume) on whose boundary S the fluid will adhere. If S_0 denotes the set of boundary points at $t = 0$, and if $\mathbf{r}_s(t, \mathbf{r}_0)$ is the position of a boundary point initially at $\mathbf{r}_0$, the condition on the fluid at the boundary is

$$\mathbf{u}(\mathbf{r}_s(t), t) = \dot{\mathbf{r}}_s(t, \mathbf{r}_0), \qquad \mathbf{r}_0 \in S_0 \tag{1.2}$$

One of the principal difficulties in the problems studied here comes from the fact that S may have a complicated time dependent form, which depends upon the *response* of the fluid to its changes of shape. This complex interaction can be avoided by tackling a simpler problem with S *prescribed;* in this case we may visualize the organism as "held fixed" and the propulsive force caused by the known motion of S as the unknown. For free swimming or flying, on the other hand, the motions of S must be found as part of the solution. As Newton's third law applies to the system of particles comprising organism and fluid, the total force exerted by the fluid upon S must equal the rate of change of linear momentum of the organism minus the gravitational force. The (time) average force acting upon a neutrally buoyant fish swimming through the water with constant average velocity, and therefore the average force exerted by the fish on the water, is zero! A swimming fish that is held, then released, will accelerate until the mean propulsive force just cancels the mean resistance it experiences. (Note that it is the resistance of the time-dependent boundary that is involved here.) A hovering insect must, of course, develop a mean lift equal to its weight. It might therefore be argued that the problem of propulsion is one of maintaining position (hovering) or else displacing certain particles of a material system relative to others (swimming), and that the dynamics is only a means to the end!

Problems of this kind will generally suggest some characteristic length L for the boundary S, as well as a typical frequency ω for the motions of S. As a consequence of these motions, the fluid will move about the organism with a characteristic speed U (e.g., U might be the mean swimming speed of a fish). In terms of these parameters, it is then convenient to rewrite (1.1 a) and (1.1 b) in dimensionless variables by defining

$$t^* = \omega t, \qquad \mathbf{r}^* = L^{-1}\mathbf{r}, \qquad \mathbf{u}^* = U^{-1}\mathbf{u} \qquad (1.3)$$
$$p^* = (\rho U^2)^{-1}p, \qquad \nabla^* = L\nabla$$

The Navier–Stokes equations (1.1 a) and (1.1 b) then become

$$\sigma\frac{\partial \mathbf{u}^*}{\partial t^*} + \mathbf{u}^* \cdot \nabla^*\mathbf{u}^* + \nabla^* p^* - \mathrm{Re}^{-1}\nabla^{*2}\mathbf{u}^* = \frac{F^{-1}\mathbf{g}}{g} \qquad (1.4)$$

$$\nabla^* \cdot \mathbf{u}^* = 0 \qquad (1.5)$$

In (1.4) there now appear three dimensionless parameters:

$$\sigma = \frac{\omega L}{U} = \text{frequency parameter} \qquad (1.6a)$$

$$F = \frac{U^2}{Lg} = \text{Froude number, } g = |\mathbf{g}| \tag{1.6b}$$

$$\text{Re} = \frac{UL}{\nu} = \text{Reynolds number} \tag{1.6c}$$

We illustrate the consequences of the dimensional similitude implied by (1.3)–(1.6) with several examples:

1. Consider two geometrically similar insects, each hovering in an infinite domain of fluid at rest at $|\mathbf{r}| = \infty$, in the presence of a gravitational field, but with $S(t)$ prescribed the same way for both. We may in each case define U to be ωL, which makes $\sigma = 1$ in (1.4). It then follows that the fields $\mathbf{u}^*(\mathbf{r}^*, t)$ and $p^*(\mathbf{r}^*, t) - (Fg)^{-1}\mathbf{g} \cdot \mathbf{r}^*$ will be the same for both provided that $L^2\omega/\nu$ is the same.

2. Consider two similar *free* hoverers with U again chosen to be ωL. We take the mass distributions within the organisms to be similar and proportional to ρ. The problems are then dynamically equivalent under an appropriate choice of units provided that Re and F are the same for each.

3. Consider two similar birds in flight with speeds U_1 and U_2, respectively, all other parameters being chosen as in (2). Equivalence then requires that all three parameters agree. Of course, from the point of view of energy required per unit time, the two systems may be far from equivalent, so on the basis of these considerations alone we cannot expect to find, in nature, dynamical equivalence over an extensive range of sizes.

Nor should we expect nature to conform to an elementary classification in terms of a few dimensionless parameters. To take just one example, we expect the Reynolds number based upon the diameter of the flagellum to be significantly different from a Reynolds number based upon some overall dimension of the organism.

With these limitations in mind, we list in Table 1.1 some characteristic parameter values.

These examples indicate that we may generally regard σ as a parameter of order 1, in some cases as a rather small parameter, whereas the Reynolds number may vary over 10 orders of magnitude in various problems of interest. [The Froude number can be absorbed into the pressure gradient, so that it plays no role in fixing the relative magnitude of terms in (1.1*a*), which is the primary issue here.]

The parameter Re is therefore overwhelmingly important in the

Table 1.1. *Estimates of the parameters σ and Re for various organisms, taking $\nu = 0.15\ cm^2/sec$ for air, $= 0.01\ cm^2/sec$ for water*

Realm	Organism	L (cm)	U (cm/sec)	ω (sec^{-1})	UL/ν = Re	$\omega L/U = \sigma$	Remarks
Stokesian realm	Bacterium	10^{-5}	10^{-2}–10^{-3}	10^4	10^{-5}	10–10^2	Limit of Navier–Stokes theory; Brownian motion affects smaller organisms
	Spermatozoan	10^{-2}–10^{-3}	10^{-2}	10^2	10^{-2}–10^{-3}	10–10^2	Flagellar diameter $\sim 10^{-5}$ cm
	Ciliated protozoan	10^{-2}	10^{-1}	10	10^{-1}	1	Cilium length $\sim 10^{-3}$ cm
Eulerian realm	Small wasp	0.06	10^2	400	15	0.25	U is wingtip speed while hovering
	Locust	4	400	20	10^4	0.2	Re appropriate to wing $\sim$ 2000
	Pigeon	25	100–1000	5	10^5	0.25	Re appropriate to wing $\sim 10^4$
	Medium-sized fish	50	100	2	5×10^4	1	

construction of models and great care is needed in choosing the scales (L and U) upon which it is based. Roughly speaking, Re measures the ratio of the forces of inertia acting on a small parcel of fluid to the force developed on the parcel by the viscous stresses. To appreciate its role, consider the swimming problem faced by the protozoan *Paramecium* (the third example in Table 1.1) from an anthropomorphic viewpoint. A human being of height 2 m, with a swimming speed of 1 m/sec, moves in water at a Reynolds number of about 2×10^6. This would be reduced to about 1 if the medium were thick syrup. To have a Reynolds number comparable to that of *Paramecium* or smaller creatures, one would have to maintain this swimming speed in a vat of warm pitch. As this comparison suggests, the fluid dynamics of flows at low and at high Reynolds numbers are significantly different, and as is clear from (1.4), the distinction arises mathematically in the relative magnitude of the viscous term in (1.4) in the two cases. The regimes Re $\ll 1$ (the Stokesian regime) and Re $\gg 1$ (the Eulerian regime) correspond roughly to the motion of microorganisms on the one hand, and of fish, insects, and birds on the other, a very convenient biological division. There are some cases, though, such as those of many of the smaller insects (the fourth example in Table 1.1), where Re is neither large nor small and yet the frequencies are so high as to make observation difficult. In studying these cases, which lie outside the range of everyday experience, some new fluid-dynamical phenomena (described in Chapter 11) were studied. These were first noticed in a biological context.

1.2 The Navier–Stokes equations

The acceleration of a "fluid particle" (roughly, a volume of fluid small in comparison to L but large compared to length scales of random molecular motions) can be computed by considering its motion from a point $\mathbf{r}$ to $\mathbf{r} + \Delta\mathbf{r}$ in time Δt. The acceleration is approximately

$$\begin{aligned} \frac{\mathbf{u}(\mathbf{r} + \Delta\mathbf{r}, t + \Delta t) - \mathbf{u}(\mathbf{r}, t)}{\Delta t} &\simeq \frac{\partial \mathbf{u}}{\partial t}(\mathbf{r}, t) + \frac{\Delta \mathbf{r}}{\Delta t} \cdot \nabla\mathbf{u}(\mathbf{r}, t) \\ &\simeq \frac{\partial \mathbf{u}}{\partial t}(\mathbf{r}, t) + \mathbf{u}(\mathbf{r}, t) \cdot \nabla\mathbf{u}(\mathbf{r}, t) \end{aligned} \tag{1.7}$$

giving the first two terms on the left of (1.1 *a*); the combination is often written $\mathbf{d}\mathbf{u}/dt$. The representation (1.7), by focusing on the velocity field (the Eulerian description of the fluid), does not explicitly exhibit the particle motions in nonstationary flow. If particle motions are of primary interest, the Lagrangian description of the fluid may be preferred. Let $\mathbf{r}(t;\, \mathbf{r}_0)$ denote the position of a fluid particle at time t, whose position at $t = 0$ was $\mathbf{r}_0$. Then clearly $\dot{\mathbf{r}}(t;\, \mathbf{r}_0) = \mathbf{u}(\mathbf{r}(t;\, \mathbf{r}_0),\, t)$ [where $(\dot{\ }) = d/dt$] and

$$\ddot{\mathbf{r}} = \frac{\partial \mathbf{u}}{\partial t} + \dot{\mathbf{r}} \cdot \nabla \mathbf{u} = \frac{d\mathbf{u}}{dt} \tag{1.8}$$

The Lagrangian theory, based upon the form (1.8) of the acceleration, leads to rather cumbersome equations for constant density flows, owing to the complicated form taken by the pressure and viscous forces. However, it is useful in certain instances: for example, in the specification of boundary conditions as in (1.2).

The terms (1.7) represent (if we multiply through by ρ) the rate of change of momentum density at a point; "inertial force" is sometimes used for minus this quantity. (We adopt a somewhat different terminology in Chapter 8.) Inertial forces predominate over forces associated with the viscosity of the fluid whenever $\mathrm{Re} \gg 1$ provided that U/L is a reasonable measure of the derivatives of $\mathbf{u}$ with respect to the spatial coordinates.

The remaining two terms on the right of (1.1a) represent pressure and viscous force (per unit mass), respectively, and they may be expressed as the divergence of a single stress tensor $\boldsymbol{\sigma} = \{\sigma_{ij}\}$:

$$\text{the } i\text{th component of } (\rho^{-1}\nabla p - \nu\nabla^2\mathbf{u}) = -\frac{1}{\rho}\frac{\partial}{\partial x_j}\sigma_{ij}$$

We adopt here the tensor summation convention; the repeated index j on the right indicates summation over all values of j. We see that [in view of (1.1b)] we may take

$$\sigma_{ij} = -p\delta_{ij} + \mu\left(\frac{\partial u_i}{\partial x_j} + \frac{\partial u_j}{\partial x_i}\right) \tag{1.9}$$

where $\mu = \rho\nu$ is the viscosity and δ_{ij} the Kronecker delta ($\delta_{ij} = 1$ if $i = j$; $= 0$, otherwise). If we consider an oriented area element dA_j, then $\sigma_{ij}\, dA_j$ is the force exerted by the stress field on the positive side of the element (determined by its normal). Try this with pressure alone.

The term $\partial u_j/\partial x_i$ in (1.9) is *not* uniquely determined by (1.1*a*) and (1.1*b*). We have, in going from (1.1*a*) and (1.1*b*) to (1.9), in effect stipulated that the stress tensor is symmetric. This is a reasonable additional condition because, for example, without it, a viscous stress might be produced in a uniformly rotating fluid, contrary to experience (see Exercise 1.1). To put this another way, any form of σ consistent with (1.1*a*) and (1.1*b*) would not lead to measurable differences in the mathematical description of the flow, but without symmetry it would be unphysical to refer to σ as a stress tensor. This situation arises because the solenoidal condition (1.1*b*) has already been used in (1.1*a*). We now turn to (1.1*b*).

Incompressibility and constant density are not quite the same thing in fluid dynamics. Seawater is essentially incompressible but varies (slightly) in density because of its varying salinity. Air, although compressible, can be treated as incompressible and of constant density when its speed U stays small compared to the speed of sound (approximately 350 m/sec). The wingtip speed of a wasp may reach 1.6 m/sec, that of a hummingbird about 8 m/sec, and flying speeds in nature are generally less than 30 m/sec. Evidently, the condition of incompressibility is satisfied in natural flying and swimming. Given that ρ is essentially constant, (1.1*b*) follows from the conservation of mass, which is written in the form

$$\frac{\partial \rho}{\partial t} + \nabla \cdot (\rho \mathbf{u}) = 0$$

To derive (1.1*a*) from Newton's second law, consider the integral over a small arbitrary volume V with boundary Σ. Equating the rate of change of linear momentum in V to the flux of momentum *into* V as well as the force applied to the boundary Σ and the body force due to gravity, we have

$$\frac{\partial}{\partial t} \int \rho \mathbf{u} \, dV = -\int \rho \mathbf{u} \, \mathbf{u} \cdot \mathbf{n} \, d\Sigma + \int \sigma \cdot \mathbf{n} \, d\Sigma + \int \rho \mathbf{g} \, dV \quad (1.10)$$

where $\mathbf{n}$ is the outward normal to Σ.† Applying the divergence theo-

†Wherever convenient, we let the differential imply the domain of integration, as in (1.10).

rem to the surface integrals in (1.9) and using (1.1*b*), it follows that (1.1*a*) is satisfied pointwise.

Equations (1.1*a*) and (1.1*b*) are invariant under Galilean transformations, a property that will be important in the placing of an observer for a description of locomotion. For a constant vector $\mathbf{U}$, let new variables be defined $t' = t$, $\mathbf{r}' = \mathbf{r} - \mathbf{U}t$, $\mathbf{u}' = \mathbf{u} - \mathbf{U}$, $p' = p$. It is easily seen that (1.1*a*) and (1.1*b*) have the same form in primed variables. Note that $\mathbf{u}'$ is the velocity of the fluid as seen by an observer moving relative to the original coordinate frame with speed $\mathbf{U}$, and that boundary conditions, such as the condition $\mathbf{u}(\infty, t) = 0$, will not in general be invariant under Galilean transformation.

1.3 Application to a moving organism

Redefining pressure in (1.1*a*) to absorb the gravitational force, we now consider the locomotion of an organism having specified boundary $S(t)$. Let us suppose that we first establish a coordinate system fixed relative to the fluid (at rest) at infinity. We refer to this as the *stationary* frame. Suppose now that a second frame, moving with constant speed $-\mathbf{U}$ relative to the first, can be defined such that relative to it there is a large fixed sphere Σ, with center at the moving origin, which contains S for all time. We refer to this second frame as the *comoving* frame and to $-\mathbf{U}$ as the mean velocity of the organism. (In particular, $-\mathbf{U}$ is the time average of the velocity of the centroid of the organism, and $\mathbf{U}$ is the "velocity at infinity" seen by the comoving observer.)

Consider, then, the momentum valance for the region V between S and Σ. We use the fact that, for any quantity Q defined in V,

$$\frac{\partial}{\partial t}\int Q\,dV = \int \frac{\partial}{\partial t} Q\,dV - \int Q\,\dot{\mathbf{r}}_s \cdot \mathbf{n}\,dS \tag{1.11}$$

where $\mathbf{n}$ is outward from S. Proceeding as in (1.10) and using (1.11) with $Q = \rho\mathbf{u}$, we have that

$$\frac{\partial}{\partial t}\int \rho\mathbf{u}\,dV = \int (-\rho\mathbf{u}\,\mathbf{u}\cdot\mathbf{n} + \sigma\cdot\mathbf{n})\,d\Sigma + \mathbf{F} \tag{1.12}$$

where

$$\mathbf{F} = \int \rho\mathbf{u}\,(\mathbf{u} - \dot{\mathbf{r}}_s)\cdot\mathbf{n}\,dS - \int \sigma\cdot\mathbf{n}\,dS = -\int \sigma\cdot\mathbf{n}\,dS$$

in view of the boundary condition (1.2); **F** is the instantaneous force exerted by the creature on the fluid. Defining the (temporal) mean value of a quantity Q by

$$\langle Q \rangle = \lim_{T \to \infty} \frac{1}{T} \int_0^T Q \, dt$$

and assuming that the total momentum in V is bounded in time, we obtain from (1.12) that

$$\langle \mathbf{F} \rangle = \int \langle \rho \mathbf{u} \, \mathbf{u} \cdot \mathbf{n} - \sigma \cdot \mathbf{n} \rangle \, d\Sigma = 0 \qquad (1.13)$$

because the mean force in free (unaccelerated) locomotion must vanish. Consequently, the mean momentum flux through Σ exactly balances the mean stress on Σ in free movement. Physically, the organism propels itself by pushing backward, acting on parts of its surface like a source of (say) positive momentum. At the same time, its resistance tends to push the medium forward relative to its motion at ∞, thus creating a source of negative momentum. When the mean motion is uniform, the two fluxes cancel on average. Condition (1.13) then implies something about the behavior of the wake region in the neighborhood of ∞.

Consider now the situation where the organism is restrained by an external force to remain near a fixed point. In this case $-\langle \mathbf{F} \rangle$ measures the (mean) force developed by the organism. Once motion occurs, we define net thrust = *thrust* − *drag* to be the component of the force of the fluid on the obstacle in the direction of its mean motion. For motion without neutral buoyancy $-\langle \mathbf{F} \rangle$ will also contain a part perpendicular to **U**, equal to the *lift* vector.

Once the organism is released, it will accelerate until $\langle \mathbf{F} \rangle = 0$. In the free state the decomposition of the equilibrium into the statements "thrust = drag" and "lift = effective weight" can be made whenever thrust and drag can be unambiguously computed. This can be done, for example, for swimming in the Stokesian realm (Chapter 2).

1.4 Energy and work

The rate of working on the surrounding fluid of a moving surface $S(t)$ is obtained by integrating the scalar product of velocity and the force on the fluid over S:

$$W_S(t) = -\int \mathbf{u} \cdot \sigma \cdot \mathbf{n}\, dS \tag{1.14}$$

To relate (1.14) to the first law of thermodynamics (identification of heat as a form of energy), we use the scalar product of **u** with (1.1 *a*), together with (1.1 *b*) and (1.9). Integrating over V (bounded by S and Σ), we have, after an integration by parts and use of (1.11) and (1.13),

$$\begin{aligned}\frac{d}{dt}\left(\frac{\rho}{2}\int u^2\, dV\right) + \frac{\rho}{2}\int u^2\mathbf{u} \cdot \mathbf{n}\, d\Sigma - W_S - \int \mathbf{u} \cdot \sigma \cdot \mathbf{n}\, d\Sigma \\ -\rho\int \mathbf{g} \cdot \mathbf{u}\, dV &= -\mu\int \frac{\partial u_i}{\partial x_j}\left(\frac{\partial u_i}{\partial x_j} + \frac{\partial u_j}{\partial x_i}\right) dV \\ &= -\frac{\mu}{2}\int \left(\frac{\partial u_i}{\partial x_j} + \frac{\partial u_j}{\partial x_i}\right)\left(\frac{\partial u_i}{\partial x_j} + \frac{\partial u_j}{\partial x_i}\right) dV \\ &\equiv -\Phi(t) \leq 0 \end{aligned} \tag{1.15}$$

We can write (1.15) in the form

$$W_S + W_\Sigma + G = \dot{E} + \mathcal{F}_\Sigma + \Phi \tag{1.16}$$

where $\dot{E}$ = rate of change of kinetic energy in V
W_Σ = rate of working of forces on S on fluid in V
G = rate of working of gravitational body force on fluid in V
$\mathcal{F}_\Sigma$ = flux (outward) of kinetic energy through Σ
Φ = rate of appearance of heat due to viscous dissipation

Thus, (1.16) expresses the first law: rate of working = change in energy (here entirely contained in the kinetic energy) plus the rate of output of heat.

Suppose now that a free organism is moving with constant mean velocity in a fluid at rest at ∞. Again, by redefining pressure, the term G can be removed from (1.16), in which case W_S and W_Σ represent the rates of working with the new pressure. However, these must be the same as the old because the gravitational field is conservative. Also, it is reasonable that the work done on Σ and the flux of kinetic energy through Σ will vanish as Σ expands to ∞. (This can be verified from the decay of the flow field in a viscous fluid.) Thus, (1.16) reduces to

$$W_S = \dot{E} + \Phi \tag{1.17}$$

taken over the whole exterior volume.

From the time average of (1.17), we see that the mean work done by the surface appears as heat generated by viscous dissipation. The case of an inviscid fluid ($\mu \equiv 0$) must be treated separately, for although $\Phi = 0$, the Σ integrals cannot be discarded. In general, work done at S produces a flux of kinetic energy at ∞, and, because of this singularity at ∞, a drag is realized without viscous dissipation (see Chapter 11).

Of course, not all dissipative losses in natural propulsion occur in the fluid. Heat produced by the metabolism of the organism has not been accounted for. To take a concrete example, a hollow rubber ball containing a device for periodically contracting the device into an ellipsoid will produce motions in the fluid and satisfy the foregoing expression for the dissipation. However, heat will also be generated within the ball, and the total work being done by the mechanism will equal the sum of the two.

A mechanical *efficiency* of flying or swimming can be defined only if a standard dissipation corresponding to optimal propulsion is defined. For example, if mean motion is with speed U and the thrust T can be defined, the quantity

$$\eta = \frac{UT}{\langle \Phi \rangle} = \frac{UT}{\langle W_S \rangle} \tag{1.18}$$

will prove useful.† A true thermodynamic efficiency would be

$$\eta' = \frac{UT}{\langle \Phi \rangle + \langle \Phi' \rangle} \tag{1.19}$$

where $\Phi'(t)$ is the heat output of the organism. In the following chapters we shall from time to time compute the efficiency of propulsion, using (1.18), without attempting to evaluate its importance as a factor in the evolution of the mechanism under study. If a significant fraction of the energy available to an organism is expended in its locomotion, and if this ability is in some way essential to its survival, we would expect natural selection to favor the higher efficiency. There is therefore usually an extremal problem associated with each propulsive strategy. But it is not obvious, and indeed is unlikely, that evolution should have invariably "solved" this extremal problem, and our atti-

†This is sometimes referred to as the Froude efficiency from its role in the theory of mechanical impellers.

tude will be to view the mechanical efficiency as one of perhaps many parameters that can be useful in the evaluation of experimental data through mechanistic models.

1.5 General references

A more formal, detailed treatment of the equations of motion may be found in many texts [see, in particular, Lamb (1932), Sommerfeld (1952), and Batchelor (1967)]. The books by Prandtl (1952) and Landau and Lifshitz (1959) are recommended highly. The problems discussed below are based largely on the excellent monograph of Lighthill (1975) and the comprehensive proceedings of the recent conference on swimming and flying in nature, edited by T. Y.-T. Wu et al (1975).

Exercises

1.1 Show that the only linear combination of velocity derivatives that vanishes in all uniformly rotating fluids has the symmetric form of the viscous stress in (1.9).

1.2 Verify that the Navier–Stokes equations (1.1*a*) and (1.1*b*) are invariant under a Galilean transformation.

1.3 Assuming that the cross-sectional area of the downward jet of air produced by a hovering insect is proportional to the square of a representative dimension L, show that the power required to hover is proportional to $L^{7/2}$. For birds the power available for flight is known to be roughly proportional to L^2. What conclusion can be drawn regarding natural hovering?

2

The Stokesian realm: Re $\ll$ 1

2.1 The Stokes equations

We shall consider the swimming of an organism when σ is a quantity of order unity and Re $\ll 1$. The scales L and U will be taken as overall organism length and mean swimming speed, respectively. If the substitution

$$p^* = \mathrm{Re}^{-1}p^\dagger + (Fg)^{-1}\mathbf{r} \cdot \mathbf{g}$$

is made in (1.4) and superscripts are dropped, the limit Re $\rightarrow 0$ may be taken, thereby reducing the system to the formal limit

$$\nabla p - \nabla^2\mathbf{u} = 0 \tag{2.1}$$

$$\nabla \cdot \mathbf{u} = 0 \tag{2.2}$$

These are the Stokes equations, and they carry with them the exact boundary condition (1.2). They describe either very viscous fluids, or very slow motions, or flow fields of small extent, any of these cases implying small Re. We have simultaneously reduced t to the role of a parameter by the condition $\sigma\mathrm{Re} = \omega L^2/\nu \ll 1$, implying that frequencies cannot be too large. Although the flow field associated with a moving boundary may be time-dependent, insofar as the dynamics of the fluid is concerned, it is moving slowly (quasi-steady). Physically, the condition Re $\ll 1$ means that momentum is instantly diffused throughout the fluid; thus, the fluid conforms with no inertial lag to the time dependence of the boundary. As organism and fluid tend to have comparable densities, it is clear that the organism's inertia is also negligible and therefore that the force on a free neutrally buoyant Stokesian swimmer is zero *at each instant.*

Equation (2.1) is linear in $\mathbf{u}$, in contrast to (1.4). This is a significant simplification for boundary-value problems with (1.2) and prescribed $S(t)$. (If S were to depend upon $\mathbf{u}$, the problem might still be nonlinear.)

2.2 Some solutions

In view of the linearity of (2.1) and (2.2), we seek solutions of the form

$$u_i = U_{ij}a_j, \qquad p = \Pi_j a_j \tag{2.3}$$

where the a_j are arbitrary constants. Define the symmetric tensor field

$$U_{ij} = \nabla^2\chi\delta_{ij} - \frac{\partial^2\chi}{\partial x_i\,\partial x_j} \tag{2.4}$$

where χ is an unknown function to be specified. Then,

$$\frac{\partial U_{ij}}{\partial x_i} = \frac{\partial}{\partial x_j}\nabla^2\chi - \nabla^2\frac{\partial\chi}{\partial x_j} = 0$$

and we see that (2.2) is satisfied for any χ. We also note that (2.1) will be satisfied provided that we define π_j in (2.3) to cancel the Laplacian of the second term on the right of (2.4), that is,

$$\Pi_j = -\frac{\partial}{\partial x_j}\nabla^2\chi \tag{2.5}$$

and provided that

$$\nabla^4\chi = 0 \tag{2.6}$$

A solution of this biharmonic equation (2.6) generates, through (2.4) and (2.5), a solution of the Stokes equations. (Not all interesting solutions can be obtained in this form; see Exercise 2.1.)

For example, take $\chi = (8\pi)^{-1}(x_1^2 + x_2^2 + x_3^2)^{1/2} = (8\pi)^{-1}r$. (The multiplicative constant is chosen to anticipate a convenient normalization.) Then

$$U_{ij} = \frac{1}{8\pi}\left(\frac{x_i x_j}{r^3} + \frac{\delta_{ij}}{r}\right), \qquad \Pi_j = \frac{1}{4\pi}\left(\frac{x_j}{r^3}\right) \tag{2.7}$$

The solution (2.7) (or rather its scalar product with any unit vector **a**, $U_{ij}a_j$) is known as a *Stokeslet*. If the solution is regarded as a distribution and substituted into (2.1), we obtain

$$\nabla\Pi_j a_j - \nabla^2 U_{ij}a_j = -\nabla^4\chi a_i = -\frac{1}{4\pi}\nabla^2(r^{-1})a_i = \delta(\mathbf{r})a_i$$

where $\delta(r)$ is the Dirac delta function.

This formula can be verified as follows. Let δ_ϵ be a positive, radially symmetric C^∞ function that vanishes outside a sphere of radius $\epsilon > 0$ about the origin, the integral of δ_ϵ being 1. We start by solving the problem

$$\nabla^2\phi = \delta_\epsilon$$

and let $\epsilon \to 0$. The limit of δ_ϵ is the Dirac function. Let $\rho > \epsilon$ be the radius of a sphere G centered at the origin. Then

$$1 = \int \delta_\epsilon \, dG = \int \nabla^2\phi \, dG = \int_{r=\rho} \frac{\partial\phi}{\partial n} \, dS \tag{2.8}$$

Since ϕ is harmonic ($\nabla^2\phi = 0$) outside the ϵ sphere, we have

$$\frac{1}{r}\frac{d^2}{dr^2}(r\phi) = 0, \qquad \phi = A + Br^{-1}$$

The condition $\phi(\infty) = 0$ requires that $A = 0$ and (2.8) then implies that $B = -1/4\pi$.

Thus, a Stokeslet gives the response of the fluid to unit impulse at the origin in the direction of $\mathbf{a}$. For an organism swimming with mean velocity $\mathbf{U}$, it can be shown that relative to a moving observer the flow field in a neighborhood of infinity has the expansion

$$\begin{aligned} \mathbf{u} &= \mathbf{U} + U_{ij}F_j(\mathbf{t}) + O(r^{-2}) \\ p &= \pi_j F_j(\mathbf{t}) + O(r^{-3}) \end{aligned} \tag{2.9}$$

where $\mathbf{F}$ is again the force that the body exerts on the fluid. In effect, (2.9) states that in a neighborhood of ∞, the body appears as a concentrated force distribution. For free swimming, (2.9) implies that $\mathbf{u} = O(r^{-2})$ as $r \to \infty$.

To verify that (2.8) is consistent with momentum balance, we note that in Stokes' flow the momentum tensor simplifies to $\{\sigma_{ij}\}$ in (1.12), so that if Σ is a sphere of radius R,

$$\begin{aligned} \int \left[pn_i - \left(\frac{\partial u_i}{\partial x_j} + \frac{\partial u_j}{\partial x_i} \right) n_j \right] d\Sigma &= \frac{1}{4\pi} \int \frac{x_i x_j}{r^4} F_j \, d\Sigma \\ &\quad + \frac{1}{2\pi} \int \frac{x_i x_j}{r^4} F_j \, d\Sigma \\ &= F_i \end{aligned}$$

where the two integrals on the right correspond to pressure and viscous stress contributions, respectively, the latter accounting for two-thirds of the force. Recall that our present formulation is dimensionless, the dimensional force being $\rho UL\nu$ times the coefficient $\mathbf{F}$ of the Stokeslet.

2.3 Flow past a sphere

As an example of a boundary-value problem, consider the stationary flow of a fluid over a sphere of radius L, with the velocity at ∞ being $\mathbf{U}$. It is clear from (2.6) that in this problem χ will be spherically symmetric. Writing

$$u_i = U_{ij}U_j$$

with U_{ij} as in (2.4), we solve (2.6) and get

$$\chi = \tfrac{1}{4}r^2 + Ar + \frac{B}{r}$$

where the first term yields the correct behavior at ∞ and A and B are to be chosen to make $\mathbf{u}$ vanish on $r = 1$. It is easy to see from (2.4) that this will be the case if $\chi''(1) = \chi'(1) = 0$. This in turn implies that $B = \frac{1}{4}$ and $A = \frac{3}{4}$. Thus, in $r > 1$,

$$\mathbf{u} = \mathbf{U} - \tfrac{3}{4}\left(\frac{\mathbf{U}\cdot\mathbf{rr}}{r^3} + \frac{\mathbf{U}}{r}\right) - \tfrac{1}{4}\nabla\left(\frac{\mathbf{U}\cdot\mathbf{r}}{r^3}\right)$$

$$p = -\tfrac{3}{2}\frac{\mathbf{U}\cdot\mathbf{r}}{r^3}$$

The term multiplied by $-\frac{3}{4}$ is a Stokeslet corresponding to the force $-\frac{3}{4}\cdot 8\pi\mathbf{U} = -6\pi\mathbf{U}$. The dimensional drag on a sphere of radius L in a flow of speed U is, accordingly, $6\pi UL\mu$. This is the famous Stokes law for a sphere. A well-known application was in Millikan's oil-drop experiment, which measured the charge of an electron; the formula is useful in the biological context as, for example, a rough indication of the drag of the head of a spermatozoan.

2.4 Time-reversal symmetry

We show here that organisms having time-reversal symmetry cannot swim in the Stokesian realm.

Consider an organism with boundary $S(t)$ swimming with mean speed $\mathbf{U}$ in the Stokesian realm. We may, relative to the comoving observer, decompose **u** as follows (we use linearity; a corresponding decomposition of the pressure is implied):

$$\begin{aligned} \mathbf{u} &= \mathbf{U} + \mathbf{u}_1 + \mathbf{u}_2 \\ \mathbf{u}_1(S) &= -\mathbf{U}, \qquad \mathbf{u}_1(\infty) = 0 \\ \mathbf{u}_2(S) &= \dot{\mathbf{r}}_s \equiv \mathbf{u}_s, \qquad \mathbf{u}_2(\infty) = 0 \end{aligned} \tag{2.10}$$

Thus, (2.3) is satisfied in two steps.† We also may write $\mathbf{u}_1 = \mathbf{B} \cdot \mathbf{U}$, where the tensor field **B** depends only on $S(t)$. The time average of (1.17) may be applied to the flow $\mathbf{u}_1$, p_1, to give

$$\langle \mathbf{F}_1 \rangle \cdot \mathbf{U} + \langle \Phi_1 \rangle = 0 \tag{2.11}$$

where $\mathbf{F}_1$ is the force on the fluid associated with this part of **u** and Φ_1 is the corresponding dissipation. Also, let $\mathbf{F}_2$ be the force on the fluid due to u_2, p_2. In free swimming we have

$$\langle \mathbf{F}_1 \rangle + \langle \mathbf{F}_2 \rangle = \mathbf{g}(M_0 - \rho V_0) \tag{2.12}$$

where M_0 is the mass and V_0 the mean volume of the organism.

Suppose now that a second organism is introduced, having a boundary motion $S^*(t) = S(-t)$; we simply run the propulsive mechanism of the original organism in reverse. We can do this by reversing the velocity of all points of S at $t = 0$, thereby repeating the same time sequence of shapes with the velocity of boundary points reversed. In particular, $\mathbf{u}_2^*(t) = -\mathbf{u}_2(t)$ and therefore $\langle \mathbf{F}_2^* \rangle = -\langle \mathbf{F}_2 \rangle$. For the component $\mathbf{u}_1^*$ we must solve the Stokes equations with **u** equal to $-\mathbf{U}$ on the boundary $S(-t)$.

Let us suppose further that the comoving observer cannot distinguish between the two organisms. This assumption of time-reversal symmetry could be implied by the indistinguishability of forward and backward running of a mechanism generating a prescribed boundary, for example. It could be tested by running forward and backward a film taken by the comoving observer. If the symmetry is found, we have $\langle \mathbf{F}_2^* \rangle = \langle \mathbf{F}_2 \rangle = -\langle \mathbf{F}_2 \rangle$, and consequently

$$\langle \mathbf{F}_1 \rangle = \mathbf{g}(M_0 - \rho V_0), \qquad \langle \mathbf{F}_2 \rangle = 0 \tag{2.13}$$

If the organism is neutrally buoyant, the right-hand side of (2.12) van-

†This decomposition assumes the existence of the constituent Stokes flows. This will be the case if S bounds a region of finite volume but is not true for the analogous two-dimensional problem.

ishes and (2.11) reduces to $\Phi_1 = 0$, implying [by way of (2.10)] that $\mathbf{u}_1$ = constant = 0 and therefore $\mathbf{U} = 0$. We thus have the following theorem: *A neutrally buoyant organism exhibiting time-reversal symmetry is a nonswimmer in the Stokesian realm.*

A few remarks concerning this intuitively appealing result are in order.

1. For organisms that are not neutrally buoyant, a trivial case of "swimming" can result from buoyancy. In fact, owing to asymmetries, an essentially "inert" form may move horizontally as well as vertically. In a sense the true test is the case of neutral buoyancy.

2. The result fails unless Re and σ Re are both small. Stokesian nonswimmers (such as the one shown in Figure 2.2*a*) may propel themselves nicely at higher Reynolds numbers. Note that the full Navier–Stokes equations are *not* invariant under $t \to -t$ or simultaneously $t \to -t$, $\mathbf{u} \to -\mathbf{u}$. The theorem is, in fact, quite misleading concerning possible propulsion at higher Reynolds numbers.

3. The theorem can actually be strengthened considerably by noting that the key point is the change of sign of the time *average* of F_2. Because of the quasi-steady nature of the dynamical problem, what matters is only the sequence of configurations, not the time at which they occur. So it is sufficient to require only "essential" time-reversal symmetry. We define $S(t)$ to be essentially time-reversable on $-\infty < t < +\infty$ provided that $S(-t)$ is indistinguishable from $S(\tau(t))$ for some $\tau(t)$ with positive derivative. That is, we require only that reversal of a film of the motion be indistinguishable from forward projection when the projectionist is allowed to vary the film speed in a certain way.† A clam shell that opens slowly and shuts quickly provides an example of a motion that is essentially time-reversible. To prove the generalized theorem, we need only note that the velocity of a point of the boundary of an essentially time reversible swimmer satisfies

$$\mathbf{u}_s^*(t) = -\mathbf{u}_s(-t) = \dot{\tau}(t)\mathbf{u}_s(\tau(t))$$

The instantaneous value of $\mathbf{F}_2$ depends linearly on $\mathbf{u}_s$ and therefore

$$\mathbf{F}_2^*(t) = -\mathbf{F}_2(-t) = \dot{\tau}(t)\mathbf{F}_2(\tau(t))$$

But then $\langle \mathbf{F}^*{}_2 \rangle = -\langle \mathbf{F}_2 \rangle = \langle \mathbf{F}_2 \rangle$, and so again $\langle \mathbf{F}_2 \rangle = 0$.

†Purcell (1977), in a stimulating discussion of Stokesian hydrodynamics, refers to these movements as *reciprocal*.

Figure 2.1. A progressive wave on an organism's surface is not time-reversible even though a boundary point moves on a straight-line segment and each orbit *is* time-reversible.

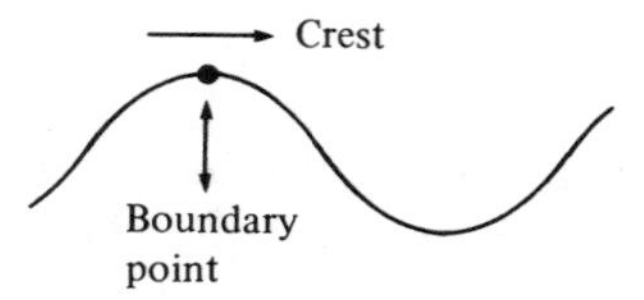

4. Further generalizations, incorporating additional symmetry of the Stokes flow problem, can be given (see Exercise 2.2).

5. The meaning of time-reversal symmetry should be carefully weighed. It is not simply a matter of material orbits of boundary points being indistinguishable, because the direction of time might be implied by the phase difference between two orbits. For example, a surface may propagate progressive waves of fixed-wave form in such a way that the material points relative to the organism move on straight-line segments (see Figure 2.1). However, the phase velocity of the wave will change sign under time reversal. Indeed, as we see in Chapters 2 and 5, propagation of waves down a boundary provides a very useful and adaptable propulsive mechanism at low Reynolds numbers.

We consider now several examples:

1. An ellipsoidal body (head) contains a mechanism that flaps a rigid fin up and down (symmetrically with respect to the head; see Figure 2.2*a*). This device, made neutrally bouyant, will not swim in syrup (but will swim in water).

2. The head now rotates a rigid helical wire, tilted into a flexible sleeve attached to the head (as the wire turns, the head turns in the opposite direction to balance the torque; see Figure 2.2*b*). With this motion, which is *not* time-reversible, a helical wave passes down the wire. It turns out that this body swims in syrup (but not well in water).

G. I. Taylor actually constructed such models and performed this delightful experiment, with the predicted results.†

†See the film "Low Reynolds Number Flows," prepared by the National Committee for Fluid Mechanics Films, A.I.A.A. Educational Programs.

2.5 Efficiency

Using linearity and the decomposition (2.10) we may separately compute the force with which the restrained organism acts on the fluid [see (1.12)]:

$$F_2 = -\int \sigma_2 \cdot \mathbf{n}\, dS$$

Because swimming with velocity $\mathbf{U}$ relative to the fluid at ∞ implies that the mean rate of working of this thrust force is $\langle \mathbf{F}_2 \cdot \mathbf{U} \rangle$, mechanical efficiency can be defined by

$$\eta = \frac{\langle \mathbf{F}_2 \cdot \mathbf{U} \rangle}{\langle \Phi \rangle} = -\frac{\langle \mathbf{F}_1 \cdot \mathbf{U} \rangle}{\langle \Phi \rangle} \tag{2.14}$$

where Φ is the dissipation of the flow $\mathbf{u}_1 + \mathbf{u}_2$. It is not difficult to show that

$$\langle \mathbf{F}_1 \cdot \mathbf{U} \rangle = -\langle \Phi_{11} \rangle = \langle \Phi_{21} \rangle \tag{2.15}$$

where

$$\Phi_{rs} = \frac{\mu}{2} \int \left(\frac{\partial u_{ri}}{\partial x_j} + \frac{\partial u_{rj}}{\partial x_i} \right) \left(\frac{\partial u_{si}}{\partial x_j} + \frac{\partial u_{sj}}{\partial x_i} \right) dV, \qquad r, s = 1, 2$$

Thus, (2.14) may be written

$$\eta = \frac{\langle \Phi_{11} \rangle}{\langle \Phi_{11} + 2\Phi_{12} + \Phi_{22} \rangle} = \frac{\langle \Phi_{11} \rangle}{\langle \Phi_{22} \rangle - \langle \Phi_{11} \rangle} \tag{2.16}$$

Figure 2.2. Fin and tail configurations. (*a*) Flapping fin. (*b*) Helical tail. (*c*) Straight tail. (*a*) and (*c*) are Stokesian nonswimmers.

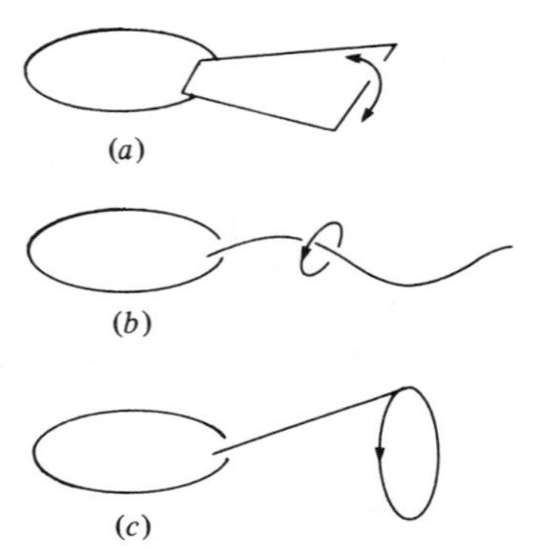

It is not obvious from (2.16) that this η is necessarily less than 1, but its physical meaning suggests that this is always the case. Thus, (2.16) raises the interesting problem of obtaining upper bounds on η for classes of deformations of S: given Φ_{11} and Φ_{22} corresponding to flows satisfying (2.10) that develop equal and opposite mean forces on S, what is the least upper bound on Φ_{11}/Φ_{22}? We shall not pursue this question here but will give some examples in subsequent chapters.

Exercises

2.1 Construct a set of solutions of the Stokes equations having the form $u_i = \epsilon_{ijk}(\partial\chi/\partial x_j)a_k$, where $\epsilon_{ijk} = \pm 1$, depending upon whether (i, j, k) is an even or odd permutation of (1, 2, 3). What is the associated pressure field? Use this form to find the flow field generated by a sphere of radius L spinning with angular velocity Ω in a fluid at rest. Show that the torque on the sphere is $8\pi\Omega L^3\mu$. (Solutions of this form, together with those considered above, provide a complete set for an exterior domain.)

2.2 Prove that an organism which moves a rigid tail so as to sweep out a cone with axis coincident with the axis of its ellipsoidal head is a nonswimmer (see Figure 2.2*c*; note that this motion is not time-reversible). Do this by showing that the mirror image of a Stokes flow is a Stokes flow, and therefore that *if time reversal of a boundary motion is indistinguishable from its mirror image, we generate a Stokesian nonswimmer.*

2.3 Establish (2.15) by integrating $\mathbf{u}_1 \cdot \sigma_1$ and $\mathbf{u}_2 \cdot \sigma_2$ over V and assuming that $\mathbf{u}_1$ and $\mathbf{u}_2$ decay at ∞ like Stokeslets.

3

Swimming of a sheet

3.1 Formulation

To take a specific example of propulsion in the Stokesian realm, we consider now the swimming of a thin flexible sheet. We first study the case of a sheet that is infinite, undergoing two-dimensional periodic deformations about an unperturbed position $y = 0$ (see Figure 3.1). Later, we shall return to some implications of the problem for finite swimmers.

The admittedly unrealistic geometry is chosen to allow a relatively straightforward solution of the Stokes equations. Nevertheless, models of this kind have an important application to ciliary propulsion (Chapter 7), and they raise a number of interesting and subtle properties of time-dependent Stokes' flows, against which we can test our physical intuition. A version of the problem was considered by G. I. Taylor (1951) in the first of a series of pioneering papers on the swimming of microorganisms. More recently, the model has been reexamined (in the context of ciliary models) by Blake (1971*a*) and Brennen (1974, 1975).

We take the sheet to be of infinitesimal thickness and restrict attention to one side. Suppose that the sheet moves in such a way that at any time the points $(x_s(t), y_s(t))$ can be expressed as a function of x alone (no overturning). The example we consider is

$$\begin{aligned} x_s &= x + a\cos(kx - \omega t - \phi) \\ &= x + \beta\cos(kx - \omega t) + \gamma\sin(kx - \omega t) \qquad (3.1a) \\ y_s &= b\sin(kx - \omega t) \qquad (3.1b) \end{aligned}$$

where $\beta = a\cos\phi$ and $\gamma = a\sin\phi$. Note that the sheet moves with a combination of a progressive wave, with phase speed ω/k, and periodic tangential movements. The simple progressive wave is obtained when $a = 0$, and the sheet stays a plane if $b = 0$. Generally, such movements require that the sheet be extensible, but of course more general expressions, involving Fourier series in $kx - \omega t$, can be used to represent an inextensible sheet (see below). We assume that $k > 0$.

Recalling (1.2), we obtain the boundary conditions on the velocity field (u, v) from (3.1):

$$u(x_s, y_s, t) = \dot{x}_s = a\omega \sin(kx - \omega t - \phi) \tag{3.2a}$$

$$v(x_s, y_s, t) = \dot{y}_s = -b\omega \cos(kx - \omega t) \tag{3.2b}$$

The two-dimensional Stokes equations are

$$\frac{\partial p}{\partial x} - \mu \nabla^2 u = 0 \tag{3.3a}$$

$$\frac{\partial p}{\partial y} - \mu \nabla^2 v = 0 \tag{3.3b}$$

$$\frac{\partial u}{\partial x} + \frac{\partial v}{\partial y} = 0 \tag{3.3c}$$

To satisfy (3.3*c*), we recall that $(u_x + v_y)\, dx\, dy = d(u\, dy - v\, dx) = 0$. This implies that there is a function ψ such that $u\, dy - v\, dx = 0$. This implies ψ such that $u\, dy - v\, dx = d\psi$, or that

$$u = \frac{\partial \psi}{\partial y} \quad \text{and} \quad v = -\frac{\partial \psi}{\partial x} \tag{3.4}$$

Taking cross derivatives of (3.3*a*) and (3.3*b*) and eliminating pressure, we have

$$\nabla^4 \psi = 0 \tag{3.5}$$

Figure 3.1. Swimming of a sheet in two dimensions. (*a*) Seen by comoving observer. (*b*) Seen by observer moving with crests.

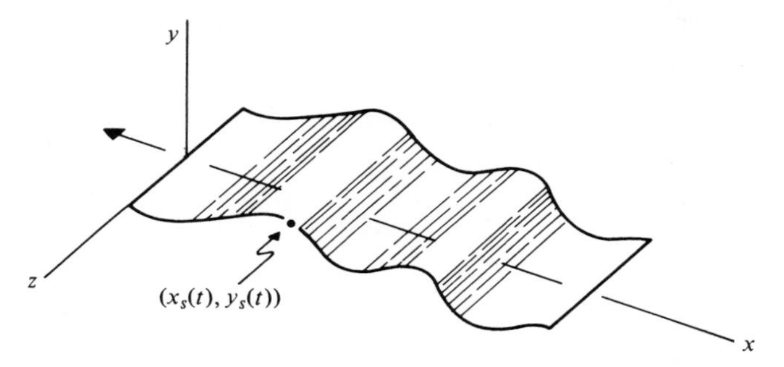

The problem is thus to find a biharmonic function ψ satisfying (3.2) and appropriate conditions at ∞. The latter are essential and we shall suppose that the sheet moves uniformly along the x axis. Relative to the present (comoving) coordinates, we have

$$\frac{\partial\psi}{\partial y} \to U = \text{constant}, \qquad \frac{\partial\psi}{\partial x} \to 0$$
$$\text{as } y \to +\infty, \quad -\infty < x < +\infty \tag{3.6}$$

The meaning of (3.6) should be noted: each material point of the sheet executes a periodic orbit according to (3.1). The constant U gives the (unknown) swimming velocity of the sheet, to be found as a function of the parameters of the orbit, and it is a consequence of the dynamical balance implicit in equations (3.3). Positive U means that the sheet swims to the left.

3.2 Expansion

The solution will be sought as a power series in bk (a and b are taken to be of comparable order). This assumption implies that the surface slope is small. It is easily checked that

$$V_n = (A_n + kB_n y)e^{-nky} \sin n(kx - \omega t) \\ + (C_n + kD_n y)e^{-nky} \cos n(kx - \omega t) \tag{3.7}$$

is a solution of (3.5) vanishing at $y = +\infty$. Another solution, compatible with (3.6), is a multiple of y. We therefore seek a solution in the form

$$\psi = \psi^{(1)} + \psi^{(2)} + \cdots, \qquad \frac{k^2}{\omega}\psi^{(m)} = O(b^m k^m) \tag{3.8}$$

We have [see (3.7)]

$$\psi^{(1)} = V_1^{(1)} + U^{(1)}y \tag{3.9a}$$

$$\psi^{(2)} = V_1^{(2)} + V_2^{(2)} + U^{(2)}y \tag{3.9b}$$

$$\vdots$$

$$\psi^{(m)} = \sum_{j=1}^{m} V_j^{(m)} + U^{(m)}y \tag{3.9c}$$

with $4m + 1$ arbitrary constants in $\psi^{(m)}$. Using (3.1) in (3.2) and expanding, we set $\xi = kx - \omega t$, $\eta = ky$, and obtain

$$\begin{aligned}
&\psi_\xi(\xi, 0) + \psi_{\xi\xi}(\xi, 0)(\beta k \cos \xi + \gamma k \sin \xi) + \psi_{\xi\eta}(\xi, 0) bk \sin \xi \\
&\quad + \tfrac{1}{2}\psi_{\xi\xi\xi}(\xi, 0)(\beta k \cos \xi + \gamma k \sin \xi)^2 + \psi_{\xi\xi\eta}(\xi, 0) bk \sin \xi \\
&\quad \times (\beta k \cos \xi + \gamma k \sin \xi) + \tfrac{1}{2}\psi_{\xi\eta\eta}(\xi, 0) b^2k^2 \sin \xi + O(b^3k^3) \\
&\qquad = \frac{\omega b}{k} \cos \xi
\end{aligned} \tag{3.10}$$

and

$$\begin{aligned}
&\psi_\eta(\xi, 0) + \psi_{\eta\xi}(\xi, 0)(\beta k \cos \xi + \gamma k \sin \xi) + \psi_{\eta\eta}(\xi, 0) bk \sin \xi \\
&\quad + \tfrac{1}{2}\psi_{\eta\xi\xi}(\xi, 0)(bk \cos \xi + \gamma k \cos \xi + \gamma k \sin \xi)^2 \\
&\quad + \psi_{\eta\xi\xi} bk \sin \xi(\beta k \cos \xi + \gamma k \sin \xi) \\
&\quad + \tfrac{1}{2}\psi_{\eta\eta\eta} b^2k^2 \sin^2 \xi + O(b^3k^3) \\
&\qquad = \frac{\beta\omega}{k} \sin \xi - \frac{\gamma\omega}{k} \cos \xi
\end{aligned} \tag{3.11}$$

At order bk, (3.7), (3.9a), (3.10), and (3.11) give

$$\begin{aligned}
\psi_\xi^{(1)}(\xi, 0) &= A_1^{(1)} \cos \xi - C_1^{(1)} \sin \xi = \frac{b\omega}{k} \cos \xi \\
\psi_\eta^{(1)}(\xi, 0) &= (B_1^{(1)} - A_1^{(1)}) \sin \xi + (D_1^{(1)} - C_1^{(1)}) \cos \xi + U^{(1)} \\
&= \frac{\beta\omega}{k} \sin \xi - \frac{\gamma\omega}{k} \cos \xi
\end{aligned}$$

Consequently, $U^{(1)} = 0$ and

$$\psi^{(1)} = \frac{\omega}{k} [(b + b\eta + \beta\eta)e^{-\eta} \sin \xi - \gamma\eta e^{-\eta} \cos \xi] \tag{3.12}$$

To order bk, the sheet does not swim. At order b^2k^2, however, (3.10) and (3.11) give

$$\begin{aligned}
\psi_\xi^{(2)}(\xi, 0) &= -\psi_{\xi\xi}^{(1)}(\xi, 0)(\beta k \cos \xi + \gamma k \sin \xi) \\
&\quad - \psi_{\xi\eta}^{(1)}(\xi, 0) bk \sin \xi
\end{aligned} \tag{3.13a}$$

$$\begin{aligned}
\psi_\eta^{(2)}(\xi, 0) &= -\psi_{\eta\xi}^{(1)}(\xi, 0)(\beta k \cos \xi + \gamma k \sin \xi) \\
&\quad - \psi_{\eta\eta}^{(1)}(\xi, 0) bk \sin \xi
\end{aligned} \tag{3.13b}$$

From (3.13b) we may deduce $U^{(2)}$ by taking the ξ average. Equation (3.12) gives

$$U \sim U^{(2)} = \tfrac{1}{2}\omega k(b^2 + 2ab\cos\phi - a^2) \tag{3.14}$$

Thus, the sheet may swim to the left or the right, depending upon the orbit. The solution at this stage is completed by noting that $V_1^{(2)} = 0$ in (3.9b) and that $A_2^{(2)} = C_2^{(2)} = 0$. Thus,

$$\psi^{(2)} = U^{(2)}y - \frac{\omega}{2}\eta e^{-2\eta}[2\gamma(b + \beta)\sin 2\xi + (b + \beta + \gamma)(b + \beta - \gamma)\cos 2\xi] \tag{3.15}$$

The pressure field corresponding to ψ may be found from the formula

$$\frac{\partial p}{\partial x} = \mu\nabla^2\frac{\partial\psi}{\partial y} \tag{3.16}$$

because the right-hand side is periodic in x with zero mean. To first order, we find

$$p^{(1)} = -2\mu\omega k e^{-\eta}[(b + \beta)\cos\xi + \gamma\sin\xi] \tag{3.17}$$

Despite the fact that the sheet does not swim with a speed of order bk, the swimming effort is dominated by the first-order contribution. The expression (1.14) reduces here to an equation for the average work per unit horizontal projected area, in the form

$$W_s(t) = \langle u(x_s, y_s, t)(\sigma_{11}n_1 + \sigma_{12}n_2) + v(x_s, y_s, t)(\sigma_{21}n_1 + \sigma_{22}n_2)\rangle \tag{3.18}$$

where $n_1(x, t)$ and $n_2(x, t)$ are the two components of the upward normal,

$$n_1 = -(1 + \Delta^2)^{-1/2}, \quad n_2 = (1 + \Delta^2)^{-1/2}, \quad \Delta = \frac{dy_s}{dx} = bk\cos\xi$$

and

$$\sigma = \begin{bmatrix} -p + 2\mu\psi_{xy} & \mu(\psi_{yy} - \psi_{xx}) \\ \mu(\psi_{yy} - \psi_{xx}) & -p - 2\mu\psi_{xy} \end{bmatrix}$$

Thus, we get

$$W_s = \mu\omega^2 k(a^2 + b^2) \tag{3.19}$$

This value is doubled if both sides of the sheet are considered.

We may interpret the swimming as an average property of the sheet's waviness. If the fluid were initially at rest at ∞ and the waves were started, the flow would build up with time until a bulk motion of the fluid "at infinity" had developed. A precise mechanical description of the origin of this bulk motion is difficult to give because it is a second-order effect. Intuitively, it is not surprising that bulk motion can be developed in a viscous fluid, but even its direction is difficult to predict from the sheet's orbit.

3.3 Inextensibility

We consider briefly Taylor's calculation, wherein the sheet is taken to be inextensible. Considerations such as this are very important for relating prescribed boundary motions to realizable deformations of an organism's surface. We consider a sheet with a sinusoidal progressive wave [$y = b \sin \xi$; this case is analogous to $a = 0$ in (3.1)] but now impose inextensibility to determine the new form of (3.1*a*). Letting $k = 1$ for simplicity, an observer moving to the right with velocity $V = \omega/k = \omega$ will see a standing wave, the material points of which move "incompressibly" along the sheet with some constant speed Q. The waveform moves with speed ω, so we have (Figure 3.2)

$$Q = \frac{V \times \text{arc length of curve in one wavelength}}{\text{wavelength}}$$

$$= \frac{\omega}{2\pi} \int_0^{2\pi} (1 + b^2 \cos^2 \xi)^{1/2} \, d\xi \tag{3.20}$$

If we now consider the boundary motion seen by our original

Figure 3.2. Progressive waves on an inextensible sheet.

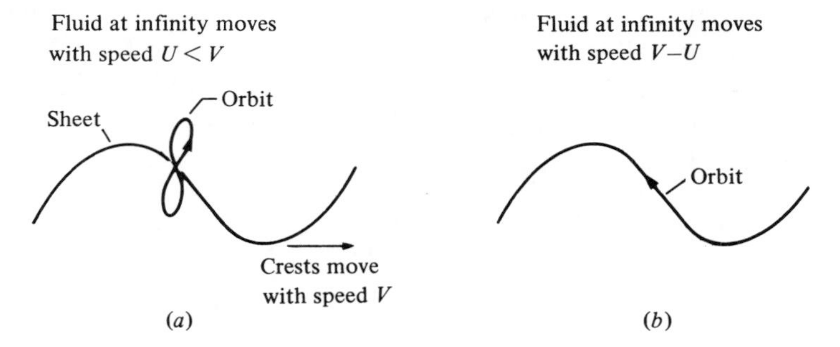

(comoving) observer, the velocity components of the boundary points are

$$u_s = -Q \cos\theta + \omega \quad \text{and} \quad v_s = -Q \sin\theta$$

where $\tan\theta = dy_s/dx = b \cos\xi$. Substituting (3.20) for Q and expanding in b, we get

$$\dot{x}_s = u_s = \frac{\omega}{4} b^2 \cos 2\xi + O(b^4)$$

$$\dot{y}_s = v_s = -\omega b \cos\xi + O(b^3)$$

These equations integrate (for small b) to

$$x_s = x - \tfrac{1}{3}b^2 \sin 2\xi \quad \text{and} \quad y_s = b \sin\xi$$

showing that material points describe figure 8's to this order.

Taylor found, by solving for ψ (as in section 3.2) for these material orbits, that

$$\frac{U}{V} = \tfrac{1}{2}b^2k^2(1 - \tfrac{19}{16}b^2k^2) + O(b^6k^6) \tag{3.21}$$

thus carrying (3.14) to fourth order. The last expression, although yielding a value of bk that minimizes U, is of course not necessarily a good approximation when $bk = O(1)$. In conclusion, it should be noted that applications of the sheet model to ciliary propulsion allow extensibility because there the sheet represents roughly a locus of cilia tips (see Chapter 7).

3.4 Effect of inertia

The present problem allows a rather simple modification to account for the effect of inertia on the first-order solution $\psi^{(1)}$, $p^{(1)}$, and it therefore provides an example of viscous propulsion spanning the full range of Reynolds number. Here the appropriate Reynolds number is denoted by $R = \omega/\nu k^2$, in terms of which the Navier–Stokes equations (in dimensionless form with k^{-1} as the unit of length and with $\omega/k^2\psi$ replacing ψ) reduce for wavelike solutions to

$$R\left(-\frac{\partial}{\partial\xi} + \psi_\eta \frac{\partial}{\partial\xi} - \psi_\xi \frac{\partial}{\partial\eta}\right)\nabla^2\psi - \nabla^4\psi = 0$$

Here R is of order 1 and fixed. Expanding ψ in powers of bk as before, we have

$$R\frac{\partial}{\partial\xi}\nabla^2\psi^{(1)} + \nabla^4\psi^{(1)} \equiv L\psi^{(1)} = 0 \qquad (3.22)$$

The appropriate solution of (3.22) [satisfying the boundary conditions at infinity is found to be

$$\psi^{(1)} = \mathcal{R}(Ae^{-\eta+i\xi} + Be^{-\lambda\eta+i\xi})$$

where $\mathcal{R}$ denotes the real part and

$$\lambda = \sqrt{1 - iR}, \qquad A = \frac{\lambda + 1}{R}(\lambda + \beta - i\gamma),$$
$$B = \frac{-(\lambda + 1)}{R}(1 + \beta - i\gamma)$$

The equation for $\psi^{(2)}$ is

$$L\psi^{(2)} = -R\left[\psi^{(1)}_\eta \frac{\partial}{\partial\xi}\nabla^2\psi^{(1)} - \psi^{(1)}_\xi \frac{\partial}{\partial_\eta}\nabla^2\psi^{(1)}\right]$$

Let the ξ average of $\psi^{(2)}$ be denoted by $\phi(\eta)$. Then the last expressions give

$$\frac{d^4}{d\eta^4}\phi = \frac{R^2}{2}\mathcal{R}[AB^*(1 + \lambda^*)e^{-(1+\lambda^*)\eta} + (\lambda + \lambda^*)|B|^2e^{-(\lambda+\lambda^*)\eta}]$$

where the star denotes complex conjugate. Integrating the last equation and applying the boundary condition at infinity, we find that

$$\phi = U^{(2)}\eta + \frac{R^2}{2}\mathcal{R}[AB^*(1 + \lambda^*)^{-3}e^{-(1+\lambda^*)\eta}$$
$$+ (\lambda + \lambda^*)^{-3}|B|^2e^{-(\lambda+\lambda^*)\eta}] \qquad (3.23)$$

From the ξ average of (3.13b) and (3.23), we then obtain

$$U^{(2)} = \mathcal{R}\{\tfrac{1}{2}[\lambda + (\lambda + 1)(\beta - i\gamma)] - \tfrac{1}{2}a^2$$
$$- \frac{R^2}{2}[AB^*(1 + \lambda^*)^{-2}(\lambda + \lambda^*)|B|^2]\} \qquad (3.24)$$

This last equation is essentially that of Tuck (1968). [See also Brennen (1974), who shows that considerable reduction is possible if new parameters for the waveform are defined.] We note here some special cases: If $a = 0$, we have that, after some reduction,

$$U^{(2)} = \tfrac{1}{2}\mathcal{R}\{\lambda + |1 + \lambda|^2[\lambda(\lambda^* + 1)^{-2} - (\lambda^* + \lambda)^{-2}]\}$$
$$= \tfrac{1}{4}\left[1 + \frac{1}{F(R)}\right]$$

where

$$F(R) = \left[\frac{1 + (1 + R^2)^{1/2}}{2}\right]^{1/2}$$

In terms of dimensional variables, the swimming speed is given by

$$\frac{U}{V} = \frac{b^2k^2}{4}\left(1 + \frac{1}{F}\right)$$

Thus, the speed is reduced by $\tfrac{1}{2}$ as R varies from 0 to ∞. It is interesting that the results appear to be uniformly valid in R, because $R = \infty$ (or $\nu = 0$) corresponds to propulsion in a fluid with zero viscosity!† If $b = 0$, the corresponding expression is

$$\frac{U}{V} = -a^2k^2\left(\frac{3F - 1}{2F}\right)$$

and again the result remains finite for large R. We note, however, that the swimming effort increases steadily with R (see Exercise 3.2).

3.5 Slowly varying waves

The sheet problem is a rather idealized case of swimming by a "finite" organism, because the infinite boundary imposes a dynamical balance from the outset. Regardless of how far the expansion of ψ is taken, it will be found that the average stress on the sheet vanishes. This is a result of the boundary condition of uniform flow at infinity, but it also rests on the extent of the surface and the fact that ψ necessarily represents the complete flow field. The terms "thrust" or "drag" cannot be used here because the requisite flow fields [determined by the decomposition (2.10)] do not exist.

†It would be of interest to examine the structure of the large R solutions separately using asymptotic methods appropriate to the simultaneous limit process $R \to \infty$, $bk \to 0$. Locomotion in an inviscid fluid without production of vorticity is considered in Chapter 8.

On the other hand, there are mass and momentum fluxes across any plane x = constant (each of which have constant time averages); moreover, if the parameters of the waveform are allowed to vary (with x, say) on a long spatial scale, the mass and momentum balance is upset on the same scale; the plate will (on average) experience a net stress and the streamlines well away from the sheet will be slightly curved.

Let f_m, f_x, and f_y denote mean fluxes of mass (relative to the fluid at infinity), x momentum, and y momentum, respectively. Then [see (3.12)].

$$\begin{aligned} f_m &= -\langle \psi(x, y_s, t) - Uy_s \rangle \\ &\sim -\langle \psi^{(1)}(x, y_s, t) \rangle = -\tfrac{1}{2}\omega\beta b \end{aligned} \tag{3.25}$$

where $\langle \cdot \rangle$ denotes either the time or x average. Similarly [see the expression for σ above (3.19); also (3.12) and (3.17)],

$$f_x = \left\langle \int_{y_s}^{\infty} \sigma_{11}\, dy \right\rangle = -2\mu\omega k\gamma b \tag{3.26a}$$

$$f_y = \left\langle \int_{y_s}^{\infty} \sigma_{21}\, dy \right\rangle = \mu\omega k\beta b \tag{3.26b}$$

Suppose now that the time average of (x_s, y_s) lies on a curved surface parameterized by arc length s. We suppose that a typical radius of curvature of this surface is large compared to k^{-1} (see Figure 3.3). Then, if a, b, ϕ, ω, and k depend on s (3.25)–(3.26b) give $f_m(s)$, $f_x(s)$, and $f_y(s)$ as slowly varying functions and the latter supplement (3.14)

Figure 3.3. Waves on a curved sheet, $L \gg k^{-1}$.

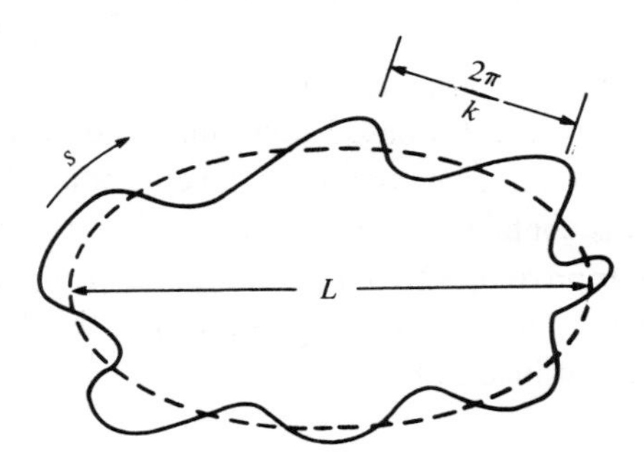

in determining the effect of the undulations of the sheet on the distant fluid. We then have the following rules for the average effect of the sheet upon the fluid well above it, relative to the comoving observer:

1. The velocity tangent to the mean surface is

$$u_t = U^{(2)}(s) \qquad \text{given by (3.14)}$$

2. The normal velocity is

$$u_n = -\frac{\partial}{\partial s} f_m(s) \tag{3.27}$$

The meaning of (3.27) is that the average increase of ψ with s corresponds to a flux of fluid into the vicinity of the sheet, an exact analog being the concept of "displacement thickness" in classical boundary-layer theory [see Batchelor (1967), p. 311]. Note that, if L is as in Figure 3.3, then $u_n \sim O(k^{-1}L^{-1})$ and is generally negligible; the kinematic effect of the sheet is primarily to induce the tangential swimming velocity $U^{(2)}$.

3. The dynamic effect of variation of sheet parameters with s is as if a tangential stress

$$\sigma_t = \frac{\partial f_x}{\partial s}$$

and a normal stress

$$\sigma_n = \frac{\partial f_y}{\partial s}$$

acted on the mean boundary of the organism. The resulting forces are to be added to those associated with the flows induced by the tangential velocity and by the displacement of the organism as a whole.

The reader may recognize that we are appealing here to a construction of the flow field of Figure 3.3 based upon matched asymptotic expansions. In effect, the oscillatory "boundary layer" induced by the movements of the sheet is confined to a layer surrounding the body of thickness $O(k^{-1})$. This approach, which has been developed (allowing for finite R) by Brennen (1974), leads to the analysis of the flow field past a ciliated organism outlined in Chapter 7.

Exercises

3.1 Using (3.21) with $bk = 0.07$, show that the predicted swimming speed corresponds to about 12 oscillations of a particle of the sheet for each wavelength of progress in the direction of swimming. (These numbers are typical of spermatozoans.)

3.2 Show that the effect of inertia on the swimming effort leads to

$$W_s = \tfrac{1}{2}\omega^2 k^2 b^2(1 + F) \qquad (a = 0)$$

$$\quad = \tfrac{1}{2}\omega^2 k^2 a^2(1 + F) \qquad (b = 0)$$

thereby generalizing (3.19) in these two cases.

3.3 Verify (3.24)–(3.26*b*).

4

The biology of low-Reynolds-number locomotion

4.1 Structure and physiology

4.1.1 *Bacteria* (references: Berg, 1975*a*, *b*; Adler, 1975)

The bacterial flagellum is very different from the organelle that bears that name in eukaryotic organisms (e.g., protozoans, algae, and multicellular organisms). It is a rigid or semirigid helical filament made of a single polymeric protein called flaggelin, which has no enzymatic activity. At the base of the filament is a complicated set of components whose structure is fairly well known by now (see Figure 4.1). The basal body serves to anchor the flagellum in the cell and probably is part of the driving motor. A number of theories of motion have been suggested, and one based on experiments of Berg, Anderson, and others now seems most probable. In this model, the flagellum is a rigid or semirigid helix that is turned at the base by a rotary motor interacting with the basal body. If true, this would apparently be the only known use of the wheel and axle for locomotion in all biology! The experiments of Berg et al. involved getting the flagellum to stick to a glass surface or to tiny latex beads. In the first case, when the flagellum was immobilized, the organism was seen to rotate; in the second case, the attached spheres could be seen to rotate. From such observations, the picture of a rigidly rotating helix has emerged.

In some species there is only one flagellum, but in others there are many, distributed in various ways over the cell. In the latter case the different flagella may rotate together as a flagellar bundle. In the cases that have been studied the overall motion of the cell is a random walk: the flagellum or flagellar bundle rotates for a certain period of time, driving the bacterium in a straight or narrowly helical "run." Then, briefly, the flagella change their sense of rotation. This is termed a "twiddle," and it may cause the cell to move backward, especially in a sufficiently viscous medium, or to "tumble" (this might be several "twiddles" interspersed with very short runs). In any case, after a bit the cell sets out again on a run in a new, presumably randomly chosen direction. When many flagella operate together in a bundle, the bundle separates during the "twiddle."

4.1.2 *Spirilla and spirochetes*

An interesting variant on the use of flagella can occur when the cell body itself is elongated and somewhat helical, as in Spirilla and Spirochetes. The position of the flagella on (or within) a spirochete is indicated in Figure 4.2. Two or more flagellar filaments are

Figure 4.1. Base of the filament of a bacterial flagellum. (After Adler, 1975.)

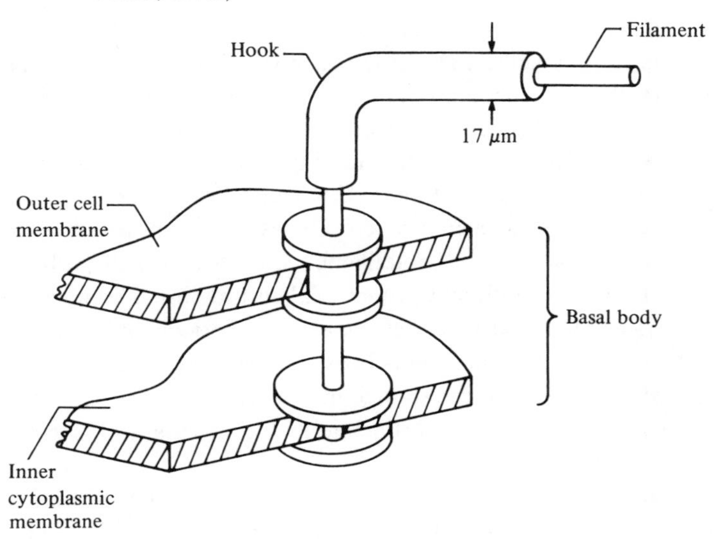

Figure 4.2. Position of a spirochete flagellum. (After Berg, 1976.)

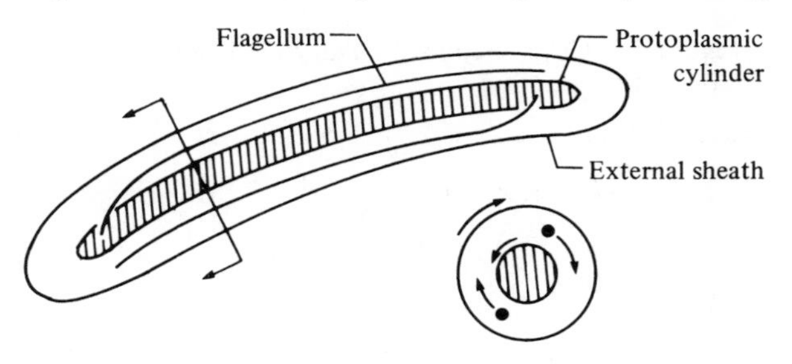

attached to the elongated, somewhat helical protoplasmic cylinder, and everything is encased in a loosely fitting, flexible external sheath. Recently, Berg (1976) has proposed that the motion of the organism may be a consequence of the rotation of the sheath. The rotation is due to the viscous stress exerted by the fluid between sheath and cylinder, set in motion by the rotation of the flagella about the cylinder. To maintain torque equilibrium the cylinder rotates in the opposite sense. The external observer sees a helical wave pass due to the movements of the semirigid body within it, down the sheath. The mechanism closely resembles the flagellum model discussed by Taylor (1952).

4.1.3 *Eucaryotic flagella* (reference: Sleigh, 1974)

These organelles have little resemblance, except in function, to bacterial flagellar filaments. They are enclosed by an extension of the cell membrane. Inside this is a structure called the *axoneme,* composed of several kinds of microtubules, linked together in the typical "9 + 2" pattern. A cross section is sketched in Figure 4.3.

The three kinds of axonemal microtubules are composed of a specific number of subunits or protofilaments, each of which probably contains at least two different proteins called tubulins. The microtubules are joined together by a delicate array of periodic linkages containing the protein dynaine, and ATPase (the enzyme involved in the extraction of energy from the storage molecule ATP); these maintain the axonemal geometry and participate in the dynamics of flagellar wave formation and propagation.

The eukaryotic flagellum is not a passive structure like the bacterial organelle; instead, it is now clear that the motile force is produced in

Figure 4.3. Cross section of eukaryotic flagellum viewed from base to tip.

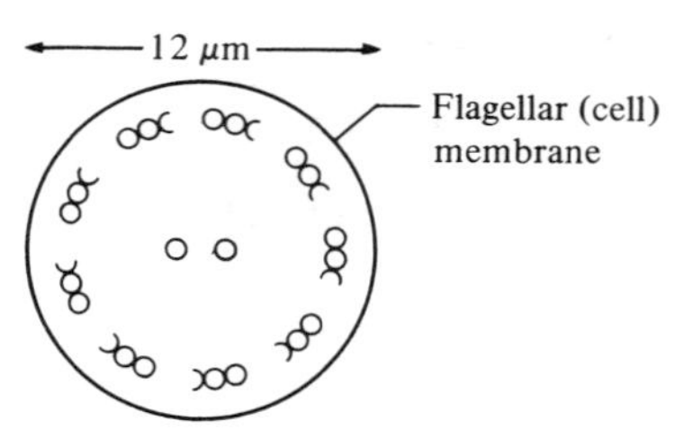

situ along the flagellum. Under appropriate conditions flagella that have been removed from the cell will continue to function autonomously, producing normal waves for some time. Some aspects of the underlying mechanism are now understood, but others remain speculative.

Peter Satir (1968) and others have shown that the propagation of flagellar waves or bends is accompanied by the sliding past one another of the peripheral axonemal doublets, which are themselves inextensible. This is apparently accomplished by the dynaine-containing connecting arms, in an unknown manner, possibly similar to the basic mechanism of muscle contraction in higher organisms.

In Satir's model it is assumed that the bending of the flagellum is a result of the relative sliding of the microtubules (Figure 4.4). Thus, a bending movement is produced that balances external and internal viscous resistance and the elastic properties of the flagellum. Once a bend has begun at the base of the flagellum, it propagates along the flagellum, followed by a successive equal bend in the opposite direction. It can be seen that the microtubules slide first one way and then another in a wavelength, so that there is no net displacement in any integral number of wavelengths. The controlling (feedback) mechanism for this pattern is not known but is thought to involve a critical deflection angle. It is worth noting that the true shape of a flagellum

Figure 4.4. Sliding microtubule model. (After Satir, 1968.)

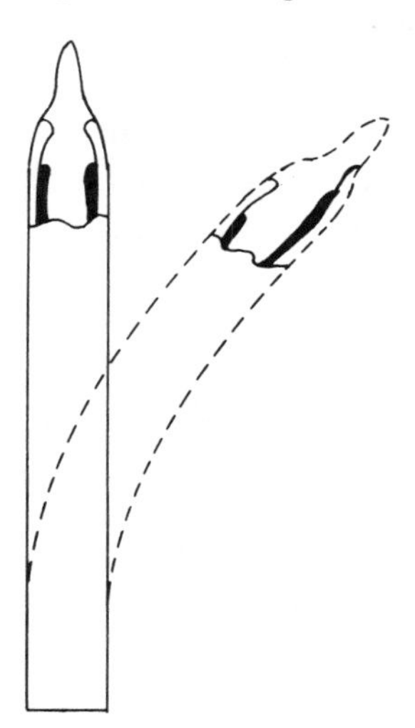

wave is not sinusoidal, but rather closer to a series of straight segments and arcs of circles. The fact will probably be important in understanding the control mechanism, because the radius of curvature of the arcs is known to vary with the viscosity of the medium.

The preceding picture of an active, energy-consuming organelle is based on work with various organisms, especially the spermatozoans of a number of invertebrate species; it is probably a good representation of the archetypal flagellum, but there are many unusual propulsive patterns, particularly among the protozoans, which differ in many ways from the above. We consider some of them below in connection with behavior.

4.1.4 *Coordination of ciliary beating*

During the last few years some very interesting work has been done on this problem, starting with the studies of Yutaka Naitoh (1974) and Roger Eckert (1972). They developed methods of measuring the electrical potential across the membrane of a protozoan, based on those used in neurophysiology. *Paramecium* is the main experimental tool.

In ciliates the basal bodies of cilia are connected just under the cell membrane by a network of microtubules called the infraciliature. It was once thought that this network served to coordinate the metachronal pattern (Chapter 7). This *neuroid* theory has now been largely abandoned because of delicate experiments in which the metachrony was not disrupted even though the infraciliature was cut. Another theory, due to Parducz (1967), held that the metachronal waves were a manifestation of spreading waves of the electrical potential across the membrane. But we now know that electric transmission in these cells is over 100 times faster than the metachronal wave speed. Most protozoans are apparently isopotential over the entire cell wall, and *Paramecium*, a very large ciliate, has an electronic decay of only a few percent over its length. Therefore, potential changes propagate rapidly, with presumably no need for action potentials or special conducting structures.

Several conclusions have emerged from these studies:

1. Detailed coordination of the beating of cilia is probably via mechanical (hydrodynamic) coupling between cilia in a region. Differences in metachrony may be due to differences in the control of

wave propagation in individual cilia. Metachrony therefore appears to be in part a fluid dynamical problem.

2. Hyperpolarizing membrane potentials induce an increase in beating frequency.

3. Depolarizing potentials increase membrane permeability to certain cations, especially Ca^{2+}, and lead to reversal of the direction of beating.

4. There are at least two components of motility; one governs cyclic ciliary bending (and is Mg^{2+}-dependent) and the other controls orientation (and is Ca^{2+}-dependent).

4.2 Behavior

4.2.1 *Spiral swimming*

The majority of swimming microorganisms appear to swim in spiral paths. In the case of bacteria and eukaryotes with helically beating flagella, the path is, to a large degree, a response of the organism as a whole to the flagellar force and torque. In ciliates the path is determined by overall body shape and the metachronal pattern of the cilia. In some ciliates (e.g., *Tetrahymena*) the organism may switch from tight, fast spiraling to open, slow spiraling. The latter path might give a way of searching or testing a wide cylindrical volume of medium.

4.2.2 *Taxes and kineses*

We consider now some aspects of sensory physiology that are linked to motility. There are two fundamental kinds of motile reaction to sensory cues. In a *kinesis* the mean *speed* but not the direction of motion changes in response to a sensory cue. This is seen clearly in the response of bacteria to the presence of gradients of various chemicals, as shown in many experiments of Adler, Berg, and others. The runs between "twiddles" are longer than average when the bacterium swims in the direction of increasing concentration of an attractant chemical, and shorter than average when it swims down the gradient. This biased random walk has the net effect of keeping the organism in the region of high concentration most of the time. Thus, the rate of diffusion, which depends upon the mean free path of the random walk (as well as the swimming speed), is affected. We may also have such a response to *level* rather than gradient of a cue, in which case under kinesis the organisms will accumulate in the region where the mean speed is slowest.

A *taxis* is a response in the direction of motion to the direction of the stimulus source. Such a response is perhaps more sophisticated and would appear to require either (1) a body large enough to separate organelles capable of sensing a gradient or (2) a memory of successive levels while in motion. It is therefore not surprising that taxes are prominent mainly among the eukaryotes.

The distinction can be seen in a simple experiment. Many protozoans are attracted by light in some band of wavelengths (see below). If one focuses a beam of light in a suspension of organisms, some species will accumulate at the focal point or some other zone of optimal intensity (kinesis), whereas others will swim toward the light and coat the side of the vessel.

4.2.3 *Specific types of cue*

Chemical. Chemosensory responses to a wide variety of chemical substances occur among most, if not all, microorganisms. Adler and others have learned a great deal about the genetic basis of the protein receptor molecules in bacterial cell walls, but as yet there is little known about coupling of receptor to the motile response. Spermatozoans of low plants and invertebrates, where studied, are found to have responses of substances (frequently proteins) associated with eggs. Honda and Miyake (1975) have demonstrated high sensitivity and specificity in responses of the ciliate *Blepharisina* to proteins called gamones, serving as sexual attractants. Hauser et al. (1975) find that the marine dinoflagellate *Cryphecodinium cohnii* has positive and negative responses to a wide variety of chemicals, including some that are neurotransmitters in higher organisms. The latter observation is intriguing in view of the association of membrane potential changes with metachronal pattern in ciliates.

Light. Many species respond to light in certain wavelengths. Recent results at the receptor level have been reviewed by Wolken (1971). In some taxes, evidence suggests that light direction is sensed by a receptor that detects the shadow of a relatively opaque structure in the cell.

Magnetism. Blakemore (1975) has reported a directional swimming response by a marine bacterium to low-strength magnetic fields. As yet, the mechanism is not understood, but it would appear to involve a curious intracellular structure of ferric material.

Gravity. True geotaxis apparently occurs in some protozoans, such as the ciliate *Tetrahymena*, but frequently the response, if observed in light, is not easily distinguished from phototaxis. The mechanism is not known, but Roberts (1970) has suggested a purely hydrodynamic model, depending upon the relative position of centers of mass and Stokes' force on a sedimenting body. A more elaborate hydrodynamic model for geotaxis in ciliates, involving in addition a coupling of the ciliary beating to the local stress field, has recently been put forward by Winet (1974).

Thigmotaxis and mechanoreception. Many benthic ciliates, especially those that live in sand ("psammophils") show a response, thigmotaxis, to surfaces. *Paramecium,* if introduced into a small glass capillary, will travel along it in a spiral with a pitch adjusted so that a certain spot on the front of the cell is always in contact with the glass.

Strickler (1975) has investigated in detail the swimming movements of small crustaceans forming a major portion of the zooplankton of ponds, lakes, and oceans. These tend to move at Reynolds numbers that are rather large for the Stokesian realm, in the range 1 to 50. One interesting aspect is the behavioral interaction of a large predator, *Cyclops*, with its prey *Colanus*, through hydrodynamical cues. The small prey species has two long, laterally projecting antennas that appear to contain many mechanoreceptors, small hairs that could in effect monitor the rates of change of the velocity components of the local flow (and therefore sense stress and vorticity). It is observed that when the predator is near, the prey stops swimming and allows itself to fall freely, the predator usually ignoring passive prey. Because of the difference in size, the hydrodynamic field of the predator can be sensed by the prey, but not the other way around.

5

Resistive-force theory of flagellar propulsion

We turn now to a model of Stokesian swimmers with flagellar tails, based upon a detailed description of the "swimming of a line." The last term is slightly misleading because it will turn out that the diameter of the flagellum enters as a parameter (although in certain results of the theory, the limit can indeed be taken). More realistically, our study will encompass swimming of organisms whose propulsive organelles are elongated and slender, with a roughly circular cross section.

The essential ingredients of the model are due to Hancock (1953) and Gray and Hancock (1955). Various aspects of the relevant hydrodynamical theory have been considered more recently by Batchelor (1970) and Cox (1970). Quite recently, a major step forward has been taken by Lighthill in an important review paper (1976), and it seems fair to say that the emerging understanding of flagellar propulsion will make it one of the triumphs of contemporary biofluiddynamics. We shall focus here on the basic Gray–Hancock model and in Chapter 6 take up the hydrodynamical problems that it raises.

5.1 Resistive-force theory

We consider first the geometrical aspects of the swimming problem using the resistive-force theory [see Lighthill (1975, pp. 53–9)]. Consider an inextensible flagellum of length L at time $t = 0$. At this instant its position can be given by parametric equations of the form

$$\mathbf{r} = (x, y, z) = \mathbf{R}(s) = (X(s), Y(s), Z(s)) \tag{5.1}$$

where s is arc length measured from one end. Then $d\mathbf{R}/ds$ is the unit tangent vector $\mathbf{t}$ in the direction of increasing s. We assume that the wave form is periodic in the sense that $X(s + \Lambda) = X(x) + \lambda$, $Y(s$

$+ \Lambda) = Y(s)$, and $Z(s + \Lambda) = Z(s)$. That is, if arc length increases by Λ, X increases by the wavelength λ. Let

$$\alpha = \frac{\lambda}{\Lambda} = \frac{\text{wavelength}}{\text{arc length of one wave}} \tag{5.2}$$

Then α is the fractional contraction in the length of the projection onto the X axis due to waviness.

Now let time increase and suppose that the wave propagates down the flagellum in the direction of increasing s (and x). Because the flagellum is inextensible, an observer moving with the phase velocity V of the wave sees material particles moving along the flagellum with velocity $-Q\mathbf{t}$ (Figure 5.1), and we have [see (3.20)]

$$V = \alpha Q \tag{5.3a}$$

This observer thus sees the material orbits:

$$\mathbf{r} = \mathbf{R}(s + Qt) \tag{5.3b}$$

and the fluid at infinity moves with velocity $(U - V)\mathbf{i}$ when $-U\mathbf{i}$ is the swimming velocity of the flagellum.

Now consider the velocity "at ∞" relative to the flagellum. Given that swimming is along the x axis, a material point on the flagellum moves relative to the fluid at infinity with velocity $-\mathbf{w}$, where

$$\mathbf{w} = (U - V)\mathbf{i} + Q\mathbf{t} \tag{5.4}$$

At this point the fundamental assumption of the resistive-force theory is made: There are constants K_T and K_N (depending upon the medium viscosity and geometrical parameters of the flagellum) such that the

Figure 5.1. The observer moves with the phase speed V, and sees the fluid at infinity moving with velocity $(U - V)\mathbf{i}$.

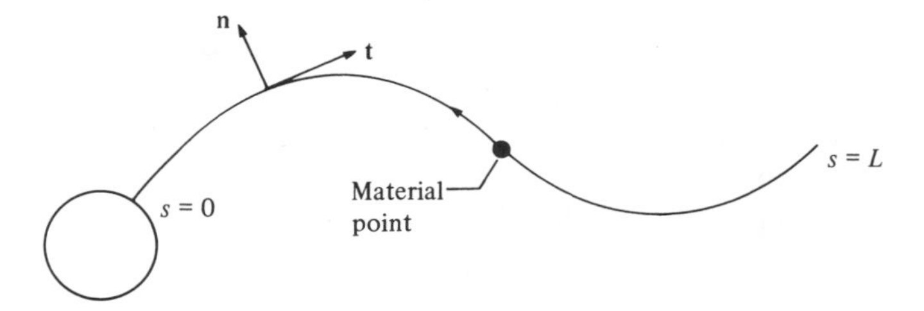

force acting on an element of flagellum of length ds equals $\mathbf{w} \cdot (K_T\mathbf{t} + K_N\mathbf{n})\, ds$. As $\mathbf{t} = d\mathbf{R}/ds$, the total force $\mathbf{F}$ may be expressed as an integral with respect to s from 0 to L:

$$\mathbf{F} = \int_0^L \mathbf{w} \cdot \mathbf{M}\, \mathbf{ds}, \qquad \mathbf{M} = (K_T - K_N)\frac{d\mathbf{R}}{ds}\frac{d\mathbf{R}}{ds} + K_N\mathbf{I} \tag{5.5}$$

From (5.4) and (5.5) it is not difficult to see that

$$F_1 = (K_T - K_N)(U - V) \int_0^L (X')^2\, ds + K_N(U - V)L + K_T Q \int_0^L X'\, ds \tag{5.6}$$

Now the net thrust T equals $-F_1$. Defining β by

$$\int_0^L (X')^2\, ds = \beta L \tag{5.7}$$

and noting from (5.3) that

$$\int_0^L X'(s)\, ds = \alpha L = \frac{V}{Q} L$$

we obtain from (5.6)

$$T = (V - U)[(K_T - K_N)\beta L + K_N L] - K_T V L \tag{5.8}$$

We may now distinguish two cases:

1. *Zero-thrust swimming* occurs if the flagellum analyzed above constitutes the entire organism. Within the resistive force theory, the appropriate value of U in (5.8) making $T = 0$ is given by

$$\frac{U_0}{V} = \frac{(\rho - 1)(\beta - 1)}{(\rho - 1)\beta + 1}, \qquad \rho = \frac{K_T}{K_N} \tag{5.9}$$

Note from (5.7) that β equals the mean-square cosine of the angle between $\mathbf{t}$ and the x axis, so that (5.9) predicts a zero-thrust swimming velocity that is opposite to the phase velocity if $0 < \rho < 1$ but in the same sense if $\rho > 1$.

2. Suppose that the flagellum propells a large inert structure, such as the head of a spermatozoan. As a result of the resistance of the

head, the thrust developed by the flagellum cannot vanish in free swimming. In fact, the thrust must equal the resistance of the head *and* the resistance of the tail due to drift relative to its zero-thrust swimming velocity. (Within the limitations of the present approach, we may simplify matters by supposing that the flow fields of head and flagellum can be considered separately, a reasonable assumption if head diameter is a small fraction of flagellum length.) As U is now the unknown and (5.7) uniquely defines the thrust of the flagellum for any U, V may in fact be found by equating T and head drag. In other words, the Gray–Hancock thrust at the actual swimming speed U may be written $-(U - V_0)[(K_T - K_N)\beta L + K_N L]$. Thus, the drag of the head is balanced by a resistive force on the flagellum developed *as it drifts backward relative to its zero-thrust swimming trajectory.* The effective resistance coefficient of the flagellum in Gray–Hancock theory, for the force opposing backward drift, is therefore $(K_T - K_N)\beta + K_N$.

To conclude, if head drag equals $LUK_N\delta$, where δ is a dimensionless coefficient, the force balance $T = LUK_N\delta$ may be solved for U/V, in terms of β, δ, and ρ, as

$$\frac{U}{V} = \frac{(\rho - 1)(\beta - 1)}{(\rho - 1)\beta + 1 + \delta} \tag{5.10}$$

Again (because of our form for head drag), K_N and K_T occur only in the ratio ρ. The typical values of ρ lie between 0.5 and 1 for low-Reynolds-number swimming. In this range, for planar or helical waves (see below), the theory agrees reasonably well with experiment. Nevertheless, what the appropriate values for ρ should be, both ideally in an infinite region and more realistically in the fluid layer of a microscope slide, is a matter not yet fully resolved.

5.2 Helical waves

These waves are relevant to observed motions of both bacterial and eukaryotic flagella, although the kinematics of the waves are fundamentally different in the two cases (Chapter 4). In addition, a new feature of the dynamical balance occurs: free swimming balances both force *and* torque.

In the eukaryotic flagellum, a superposition of two orthogonal bending waves produces a helical wave. Relative to the observer seeing tangential motion, we have

$$x = \alpha(s + Qt), \quad y = b \cos k(s + Qt), \quad z = b \sin k(s + Qt) \tag{5.11}$$

where $\alpha Q = V$ as in (5.3a) and $\alpha^2 + b^2k^2 = 1$ to obtain inextensibility. On the other hand, relative to the comoving observer (fixed to the average position of the flagellum) the velocity of a material point is

$$(\dot{x} - V, \dot{y}, \dot{z}) = (0, -bkQ \sin k(s + Qt), bkQ \cos k(s + Qt))$$

and is therefore equivalent to a rigid rotation of the helical structure about its axis. Although the two descriptions are identical for a curve, this is not so for a finite cylinder. A rigid rotation of such a structure simultaneously rotates the surface of the cross section. The distinction between eukaryotic and bacterial flagella is essentially that surface rotation accompanies the helical wave in the latter case, a fact that can have an important bearing on the torque balance. A simple laboratory model capable of passing a wave without surface rotation was constructed by Taylor (1952) by rotating a rigid helical wire in a rubber sleeve, the latter being attached rigidly to the head. (For application of the Gray–Hancock theory to helical waves, see Exercise 5.2.)

5.3 Efficiency

The rate of working of the forces exerted on the fluid is the integral of the dot product of the velocity of a material element of flagellum [$-\mathbf{w}$, where $\mathbf{w}$ is given by (5.4)] and the force with which it acts on the fluid [minus the force of resistance as given by the integrand in (5.5)]. We thus get

$$\begin{aligned} W_s &= \int_0^L [(K_T - K_N)(\mathbf{w} \cdot \mathbf{t})^2 + K_N w^2] \, ds \\ &= K_T L[(V - U)^2\beta - 2(V - U)Q\alpha + Q^2] \\ &\quad + K_N L(V - U)^2(1 - \beta) \end{aligned} \tag{5.12a}$$

It is revealing to rewrite this so as to exhibit the form of the equation $W_s = \langle\phi\rangle = \langle\Phi_{11}\rangle + 2\langle\Phi_{12}\rangle + \langle\Phi_{22}\rangle$, which entered into equation (1.17). Because [as we saw above prior to equation (5.10)] the resistance coefficient of the undulating flagellum to motions different from its zero thrust velocity is $(K_T - K_N)\beta + K_N$, the rate of working Φ_{11} may be identified with the work done moving the force $UL[(K_T - K_N)\beta + K_N]$ at speed U. On the other hand, Φ_{22} must be the rate of working of the flagellum when its swimming speed *vanishes*. Thus [using $Q\alpha = V$ in (5.12*a*) and dropping the notation for the average in the Φ's],

$$\Phi_{11} = U^2L[(K_T - K_N)\beta + K_N]$$
$$\Phi_{12} = -L(K_T - K_N)(\beta - 1)UV$$
$$\Phi_{22} = V^2K_TL(\beta - 2 + \alpha^{-2}) + K_NLV^2(1 - \beta) \qquad (5.12b)$$

It may be verified that Φ_{12} is indeed equal to $-\Phi_{11}$ *provided* that U has its zero-thrust value U_0 given by (5.9), as was required by our more general arguments.

Lighthill (1975, p. 59) introduces an efficiency (for the headless flagellum) equal to K_TLU^2/W_s. Using the Schwarz inequality to obtain the estimate $\alpha^2 \le \beta$, he concludes that the maximum efficiency occurs when $\alpha^2 = \beta$ almost everywhere (i.e., $X' =$ constant almost everywhere). The value of β maximizing η is easily seen to be $= (1 + \rho^{1/2})^{-1}$, giving

$$\eta_{\max} = (1 - \rho^{1/2})^2 \qquad (5.13)$$

and corresponding to $U/V = (1 - \rho^{1/2})^2$. For helical waves with $\beta = \cos^2\psi$ ($\psi =$ pitch angle of the helix), this gives $\psi = 42°$ for $\rho = 0.7$, a typical value even when a head is allowed.

But the most appropriate definition of efficiency is not obvious, and the considerations of Chapter 2 [see equation 2.16)] leads us to suggest one different from the above, namely

$$\eta = \frac{\Phi_{11}}{\Phi_{22} - \Phi_{11}}$$

which is maximized when Φ_{22}/Φ_{11} is minimized. The expression for this last quantity follows from (5.12*b*) and it is readily seen that again $\alpha^2 = \beta$ almost everywhere at maximum efficiency. With this choice and use of (5.9), we have

$$\frac{\Phi_{22}}{\Phi_{11}} = \frac{[(\rho - 1)\beta + 1][(\rho - 1)\beta - \rho]}{\beta(\rho - 1)^2(\beta - 1)}$$

The minimum occurs when $\beta = \frac{1}{2}$ ($\psi = 45°$), where $\Phi_{22}/\Phi_{11} = (\rho + 1)^2(\rho - 1)^{-2}$, so that

$$\eta_{\max} = \frac{(\rho - 1)^2}{4\rho} \tag{5.14}$$

We note that for $\rho = 0.7$ (5.13) and (5.14) give values of 0.027 and 0.032, respectively, so by either definition efficiency is quite low. The higher value from (5.14) is reasonable due to the fact that the comparison resistance is higher, because it refers to the resistance of the "coiled" rather than the "straightened" organism.

Of course, (5.14) would be inappropriate if it could be shown that values of ρ could be developed in Stokesian swimming, which would make the right-hand side exceed unity. This occurs when $\rho < 3 - 2\sqrt{2} = 0.172$ or $\rho > 3 + 2\sqrt{2} = 5.828$. As we shall see in Chapter 6, such values lie well outside anything predicted by asymptotic theories of flagellar hydrodynamics. It might be conjectured that the foregoing bounds provide theoretical limits on ρ, but this conclusion hinges on the correctness of using local resistance coefficients. We shall see in Chapter 6 that this ad hoc procedure is not completely justified by Stokes-flow hydrodynamics.

5.4 Remarks

We should note that the formalism of the Gray–Hancock theory can be applied to flagellar propulsion at any Reynolds number where one is prepared to accept "localness" of the resistance coefficients. Indeed, G. I. Taylor took this point of view in his 1952 paper. At higher Reynolds numbers the resistance is no longer linear in the velocity, so the thrust equation (5.8) is no longer valid. The resistance tensor at higher Reynolds must be constructed empirically from force data on cylinders of various roughnesses inclined with respect to an oncoming stream. This general class of theories provide a coherent body of propulsion mechanisms termed "anguilliform" (Lighthill, 1975, chap. 2).

If $\rho > 1$ in (5.9), the flagellum swims in the direction of propagation of the wave. Such motion, which depends on rather large *tangential* resistance, apparently occurs in nature. Examples at moderate Reynolds numbers are the marine polychaete worms, such as *Nereis*, which have large K_T because of rows of appendages on either side of a long, slender body. At low Reynolds numbers a number of protozoans, flagellates such as the Chrysomonad algae and the parasitic trypanosomatids, as well as zoospores of both brown alga and the chytid water molds, all have "flimmer" or mastigoneme flagella, with rows of tiny projecting hairs. In all these cases, locomotion is in the direction of wave propagation.

The arguments used here to compute force and moment, and hence swimming speed, have neglected what might be termed "end effects." For example, owing to the finiteness of a real flagellum, there will be hydrodynamic interaction from the cell body as well as with the fluid near the flagellum tip. In general, these will give fluctuating lateral forces as well as bending moments about axes orthogonal to the swimming direction. These will time-average to zero, but the trajectory (of the organism's centroid) is not exactly a straight line. Recently, Keller and Rubinow (1976) have analyzed in detail the path of a flagellum and cell body. For helical waves they found that the path of the organism is a helix with small radius, the axis of the flagellum being inclined to the direction of mean motion and precessing around it. For planar waves the organism moves in a wavy line. In most cases a point on the flagellum moves in a narrow figure 8.

As we have noted above, the idea of surface or *body rotation* of the cylindrical flagellum about its curved centerline provides a strategy for achieving torque balance in free swimming with helical waves. This appealing idea, due to Chwang and Wu (1971), is rather easily incorporated into the Gray–Hancock model. A circular cylinder of radius a spinning on its axis with angular velocity $\mathbf{\Omega}$ in a viscous fluid generates a simple velocity field

$$\mathbf{u} = \frac{a^2\mathbf{\Omega} \times \mathbf{r}}{r^2}$$

where r = polar radius to axis. The resulting torque per unit length acting on the cylinder is $-2\pi\mu a^2\mathbf{\Omega}$. For a curve of projected length αL, the net torque component parallel to the swimming direction is then $-2\pi\mu a^2 L\alpha\mathbf{\Omega}$, and this is available for canceling the torque generated by the wave according to the resistive force theory. The flow due to body rotation does not affect the thrust and moment calculations given earlier.

Exercises

5.1 Verify (5.6) from (5.4) and (5.5).

5.2 Let the body motion be (5.11) plus a rigid rotation with angular velocity $\Omega\mathbf{i}$. The velocity $\mathbf{w}$ is given by (5.4) plus $(0, \Omega b \sin(ks + Qt), -\Omega b \cos(ks + Qt)$. Then $\beta = \alpha^2$ and $T = (\alpha L\Omega k^{-1} - VL)[\alpha^2(K_N - K_T) + K_T] - UL[\alpha^2(K_T - K_N) + K_N]$. The torque on the flagellum (moment of the resistive force) has the x component M, where

$$\frac{\alpha M}{kb^2} = (VL - \alpha L\Omega k^{-1})[\alpha^2(K_N - K_T) + K_T] + LU\alpha^2(K_T - K_N)$$

Derive these results. From the last two we may conclude that *simultaneous force and torque equilibrium is impossible for an isolated flagellum with helical wave in the Gray–Hancock theory.* Torque balance can be established if there is a passive resisting structure (e.g., the cell body) or if the flagellar surface rotates about its curved centerline (the body rotation noted above).

5.3 Extend (5.14) to the case of a flagellum attached to a head with resistance coefficient $K_N\delta$, making use of the swimming speed given by (5.10).

6

Analysis of the flagellum

In this chapter we turn to a more formal treatment of flagellar hydrodynamics, based upon the appropriate boundary-value problem for the quasi-steady Stokes equations. To simplify the analysis, we shall, for the most part, consider a flagellum that extends to infinity in both directions. Because an *inert* body of this extent will experience an infinite resistance to translation, we will be forced to consider simultaneously the *zero-thrust* flagellum (per unit length, i.e., free swimming with zero mean force). The study will nonetheless provide a basis of comparison with the resistive-force model of Chapter 5, and will clarify the computation of the requisite resistance coefficients K_T and K_N. Later we consider briefly the hydrodynamics of a flagellum of finite length.

6.1 Local analysis

Consider first the kind of representation that might be appropriate for a long, slender body. Let $\mathbf{S}(\mathbf{r})$ denote the tensor Stokeslet field (we do not give the corresponding pressure fields in the flows that follow):

$$\mathbf{S} = (8\pi\mu)^{-1}(r^{-3}\mathbf{rr} + r^{-1}\mathbf{I}) \tag{6.1}$$

Thus, $\mathbf{S} \cdot \mathbf{f}$ is the velocity field generated by a force $\mathbf{f}\delta(\mathbf{r})$ applied to the fluid. Suppose that such forces are distributed uniformly along a short line segment. If the segment is taken to be straight and coincident with $-b < x < c$, $y = z = 0$ $(b, c > 0)$, the resulting flow field may be written as a sum of a Stokeslet field $\mathbf{u}_{st}$ associated with the tangential component $\mathbf{f}_t = f_x\mathbf{i}$, and a field $\mathbf{u}_{sn}$ generated by the orthogonal part $\mathbf{f}_n$. Thus,

$$\mathbf{u}_{st}(\mathbf{r}) = \int_{-b}^{c} \mathbf{S}(\mathbf{r} - \mathbf{R}(s)) \cdot \mathbf{f}_t \, ds \tag{6.2a}$$

$$\mathbf{u}_{sn}\mathbf{r} = \int_{-b}^{c} S(\mathbf{r} - \mathbf{R}(s)) \cdot \mathbf{f}_n \, ds \tag{6.2b}$$

where now $\mathbf{R}(s) = s\mathbf{i}$. Because a flagellum exerts a force on the fluid and can generally be regarded as thin relative to its local radius of curvature, it is likely that distributions of Stokeslets of the form (6.2) will occur in any movement. If we take the surface of the flagellum to be the envelope of spheres of radius a and center on a certain curved line (the flagellum axis), then for the segment in question, we evaluate $\mathbf{u}_{st}$ and $\mathbf{u}_{sn}$ on the circle C: $y^2 + z^2 = a$, $x = 0$, to see what surface motions there are compatible with the constant distribution of forces. If $a \ll b$, c the integrals in (6.2) can be readily expanded in a/b, a/c and one gets [see Lighthill (1975, pp. 49–51)]

$$\mathbf{u}_{st}(C) = \left[\ln\left(\frac{4bc}{a^2}\right) - 1 \right] \mathbf{f}_t + O(\epsilon^2)$$

$$\mathbf{u}_{sn}(C) = (8\pi\mu)^{-1} \ln\left(\frac{4bc}{a^2}\right) \mathbf{f}_n \tag{6.3a}$$

$$+ (4\pi\mu a^2)^{-1}\mathbf{r}(\mathbf{f}_n \cdot \mathbf{r}) + O(\epsilon^2) \tag{6.3b}$$

where $\epsilon^2 = a^2(b^2 + c^2)^{-2}$.† Because of the second term on the right of (6.3b), the Stokeslet distribution alone cannot make $\mathbf{u}$ a *constant* vector on C.

To remove the nonconstant term, consider the *source dipole* solution:

$$\mathbf{D} = (4\pi)^{-1}\nabla\nabla(r^{-1})$$

[generated by the choice $\psi = (4\pi r)^{-1}$ in (2.4)]. We distribute dipoles of moment $a^2\mathbf{f}_n/4_\mu$ on the segment,

$$\mathbf{u}_d = \frac{a^2}{4_\mu} \int_{-b}^{c} \mathbf{D}(\mathbf{r} - \mathbf{R}(s)) \cdot \mathbf{f}_n \, ds$$

to obtain on C the contribution

$$\mathbf{u}_d(C) = (8\pi\mu)^{-1}\mathbf{f}_n - (4\pi\mu a^2)^{-1}\mathbf{r}(\mathbf{f}_n \cdot \mathbf{r}) + O(\epsilon^2) \tag{6.3c}$$

Adding (6.3), we see that

$$\mathbf{u}(C) = (4\pi\mu)^{-1} \left[\ln\left(\frac{4bc}{a^2}\right) - 1 \right] \mathbf{f}_t + (8\pi\mu)^{-1} \left[\ln\left(\frac{4bc}{a^2}\right) + 1 \right] \mathbf{f}_n + O(\epsilon^2) \tag{6.4}$$

†We use the $O(\cdot)$ symbol in this chapter, as elsewhere, in its technical sense: $f(\epsilon) = O(g(\epsilon))$ if there are constants A and ϵ_0 such that $|f| < A|g|$ for $0 < \epsilon < \epsilon_0$.

From the point of view of resistive-force theory, (6.4) provides values for K_T and K_N:

$$K_T = \frac{4\pi\mu}{\ln(4bc/a^2) - 1} \quad \text{and} \quad K_N = \frac{8\pi\mu}{\ln(4bc/a^2) + 1} \tag{6.5}$$

This is essentially the procedure used by Gray and Hancock (1955) to localize the velocity-force relation as needed in the resistance-force model. The approach requires, of course, a choice of the length $(bc)^{1/2}$ in (6.5). It is reasonable to assume that this length must be determined by the wavelength of undulations of the flagellum. Lighthill (1975, p. 56) has argued that, if $\mathbf{f}$ is replaced by $\mathbf{f}e^{iks}$, where $k = 2\pi/\lambda$, $\lambda \gg a$, the integrals produce $2K_0(ka)$ wherever $\ln(4bc/a^2)$ occurred above, K_0 being the modified Bessel function of order zero. Expanding the latter for small ka, $K_0(ka) \sim \ln(2/ka - \gamma) + O(1)$, where $\gamma = 0.577\ldots$ is Euler's constant. Thus for agreement with (6.4), we should take $(bc)^{1/2} = q = \lambda/2\pi\epsilon^\gamma = 0.09\lambda$. On the other hand, Shack et al. (1974) have analyzed a helical wave using the theory of Cox (1970), and they conclude that

$$K_T = \frac{2\pi\mu}{\ln(2q/a)} \quad \text{and} \quad K_N = \frac{4\pi\mu}{\ln(2q/a) + \frac{1}{2}} \tag{6.6}$$

where q is the same constant defined above. Thus, (6.6) is obtained from (6.5) only if the -1 from the expression for K_T is dropped.

These results thus reveal the following difficulties with resistance coefficients calculated from a straight segment of flagellum: (1) the length $(bc)^{1/2}$ is unknown, and might, for example, have different values for tangential and normal Stokeslets; (2) the resistance cannot always be local and given by (6.5), because if it *were,* the helical wave would give values in agreement with (6.4) for *some* $(bc)^{1/2}$; and (3) the theory does not suggest how modifications should be made to correct for the waveform.

6.2 Formal theory

Batchelor (1970) and Cox (1970) have shown that the Stokes flow past slender bodies of characteristic length L and radius a can usually be expressed for small $\epsilon = a/L$ as asymptotic expansions in powers of $(-\ln \epsilon)^{-1}$. Although such an expansion is of fundamental importance in the study of finite flagella (see below), it is not well adapted to analysis of the zero-thrust problem. This point has recently been emphasized by Lighthill (1976), who notes that for the zero-thrust flagellum there is a considerable advantage in approaching the problem in the manner of the local analysis given above, by asking for the *boundary velocity compatible with a given force distribution* rather than the other way around (the latter procedure being, however, natural for, say, computing resistance of a finite slender body). We turn now to the formulation of Lighthill's version of the key theorem and give a proof using elementary estimates.

Again considering an infinite flagellum having a circular cross section of radius a, with axis along the curve $\mathcal{F}$: $\mathbf{r} = \mathbf{R}(s)$, the principal result of this line of investigation closely follows the local construction, and may be stated as follows:

Theorem. Let $\mathbf{u}$ be the flow field consisting of a Stokeslet distribution of strength $\mathbf{f}(s)$ on the curve $\mathcal{F}$, plus a source dipole distribution with moment $a^2\mathbf{f}_n/4\mu$ on $\mathcal{F}$, where $\mathbf{f}_n(s)$ is the projection of $\mathbf{f}$ onto the plane $\mathcal{P}(s)$ perpendicular to $\mathcal{F}$. Consider a station $s = s_0$ and let C_0 denote the circular intersection of the surface of the flagellum with $\mathcal{P}(s_0)$. Assume (1) that the improper integral

$$\int_{|s|>A} \mathbf{S}(\mathbf{r} - \mathbf{R}(s)) \cdot \mathbf{f}(s)\, ds \tag{6.7}$$

defines a function of $\mathbf{r}$, differentiable in some neighborhood of the center of C_0 for each $A > 0$; and (2) that $\mathbf{f}'(s)$ and $\mathbf{R}''(s)$ are continuous functions, uniformly for $-\infty < s < \infty$; and (3) there is a constant B such that $|\mathbf{R}(s) - \mathbf{r}_0| > B|s - s_0|$ for sufficiently large $|s - s_0|$, where $\mathbf{r}_0$ is the position vector for the center of C_0.

Let

$$f_0 = \max_s |\mathbf{f}|, \quad L^{-1} = \max_s |\mathbf{R}''(s)|, \quad l^{-1} = f_0^{-1} \max_s |\mathbf{f}'(s)| \tag{6.8}$$

and set $\epsilon = a/L$. Then, as $\epsilon \to 0$,

$$\mathbf{u}(C_0) = (4\pi\mu)^{-1}\mathbf{f}_n(s_0) + \int_{\rho>\delta} \mathbf{S}(\rho) \cdot \mathbf{f}(s)\, ds + O(E) \tag{6.9}$$

where

$$\rho(s) = \mathbf{r}_0 - \mathbf{R}(s),\ \delta = \tfrac{1}{2}a\sqrt{e}, \text{ and } E = (f_0/\mu)(1 + L/l)\epsilon^{1/2}$$

Proof of the theorem. The demonstration is straightforward, but involves a number of estimates. Let us first note that if $b = a\epsilon^{\gamma-1}$ for some γ between 0 and 1, (6.9) may be written

$$\mathbf{u}(C_0) = (4\pi\mu)^{-1}\mathbf{f}_n(s_0) + \int_{\delta<|s-s_0|<b} \mathbf{S}((s-s_0)\mathbf{t}_0)\cdot\mathbf{f}(s_0)\,ds$$
$$+ \int_{|s-s_0|>b} \mathbf{S}(\rho)\cdot\mathbf{f}(s)\,ds + O(E) + O(E\epsilon^{\gamma-1/2}) \qquad (6.10)$$

where $\mathbf{t}_0 = \mathbf{R}'(s_0)$ is the unit tangent vector at s_0. To show this, note that $\rho = \mathbf{r}_0 - \mathbf{R}(s) = \mathbf{t}_0(s-s_0) - \frac{1}{2}\mathbf{g}(s-s_0)^2$, where, by the mean value theorem, $\mathbf{g} = (X''(s_1), Y''(s_2), Z''(s_3))$ with $s_0 < s_1, s_2, s_3 < s$. Now let $\mathcal{S}(\theta) = \mathbf{S}(\mathbf{t}_0(s-s_0) + (\theta/2)\mathbf{g}(s-s_0)^2)$. Then we have, by the vector mean value theorem,

$$\mathcal{S}(1) - \mathcal{S}(0) = \nabla_r\mathcal{S}^* \cdot \tfrac{1}{2}\mathbf{g}(s-s_0)^2 = \mathbf{G}(s)$$

where the star denotes that each component of $\mathcal{S}$ is evaluated at (generally different) points θ^*, where $0 < \theta^* < 1$. But we note that because [by assumption (2) above] $|\mathbf{g}| < \sqrt{3}L^{-1}$,

$$\left|\mathbf{t}_0(s-s_0) + \frac{\theta^*}{2}\mathbf{g}(s-s_0)^2\right| \geq |s-s_0| - \frac{\sqrt{3}}{2}L^{-1}(s-s_0)^2$$

and therefore, recalling (6.1), we see that there is a constant K such that

$$\left|\int_{\delta<|s-s_0|<b} \mathbf{G}(s)\cdot\mathbf{f}(s)\,ds\right| \leq \frac{K}{8\pi}\frac{f_0 b}{\mu L}\int_{-1}^{+1} \frac{u^2}{u - (\sqrt{3}/2)(b/L)u^2}\,du$$
$$= O\left(\frac{f_0}{\mu}\epsilon^\gamma\right)$$

which would establish the result if $\mathbf{f}(s)$ appeared instead of $\mathbf{f}(s_0)$ on the right of (6.10). It remains to be shown that this change can be made, but this follows immediately from assumption (2) and the definition of l given above:

$$\left|\int_{\delta<|s-s_0|<b} \mathbf{S}(\mathbf{t}_0(s-s_0))\cdot[\mathbf{f}(s) - \mathbf{f}(s_0)]\,ds\right| = O\left(\frac{L}{l}\frac{f_0}{\mu}\epsilon^\gamma\right)$$

and therefore (6.9) and (6.10) are equivalent for any such γ.

We now wish to compare the terms on the right of (6.10) with a similar decomposition of the left-hand side. Let

$$H_1 = \left| \int_{|s-s_0|>b} [\mathbf{S}(\mathbf{r}_c - \mathbf{R}) \cdot \mathbf{f} + \frac{a^2}{4\mu} \mathbf{D}(\mathbf{r}_c - \mathbf{R}) \cdot \mathbf{f}_n] \, ds \right.$$
$$\left. - \int_{|s-s_0|>b} \mathbf{S}(\rho) \cdot \mathbf{f}(s) \, ds \right| \tag{6.11}$$

$$H_2 = \left| \int_{s_0-b}^{s_0+b} [\mathbf{S}(\mathbf{r}_c - \mathbf{R}) \cdot \mathbf{f} + \frac{a^2}{4\mu} \mathbf{D}(\mathbf{r}_c - \mathbf{R}) \cdot \mathbf{f}_n] \, ds - \frac{\mathbf{f}_n(s_0)}{4\pi\mu} \right.$$
$$\left. - \int_{\delta<|s-s_0|<b} \mathbf{S}(t_0(s - s_0)) \cdot \mathbf{f}(s_0) \, ds \right| \tag{6.12}$$

where $\mathbf{r}_c$ is any point on C_0. We first estimate H_1 using assumptions (1) and (3) above. By the vector mean value theorem, we estimate $\mathbf{S}(\mathbf{r}_c - \mathbf{R}) - \mathbf{S}(\rho)$ in terms of derivatives of the components of $\mathbf{S}$ evaluated at $r^* + \rho$, where $\mathbf{r}^*$ is a point on the line joining $\mathbf{r}_0$ and $\mathbf{r}_c$. Thus, the Stokeslet terms in (6.11) are bounded by a constant multiple of

$$\frac{f_0 a}{\mu} \int_b^{\infty} \frac{du}{(B-a)^2} = O\left(\frac{f_0}{\mu} \epsilon^{1-\gamma} \right)$$

where (3) has been used with ϵ sufficiently small. In a similar way one proves that the dipole term in (6.11) is $O[(f_0/\mu)\epsilon^{2-2\gamma}]$ and therefore

$$H_1 = O\left(\frac{f_0}{\mu} \epsilon^{1-\gamma} \right) \tag{6.13}$$

Next, we consider H_2. The idea will be to express the first integral on the right of (6.12) in a form to which the local analysis described earlier may be applied. In effect, we must replace $\mathbf{f}(s)$ by $\mathbf{f}(s_0)$ and $\mathbf{r}_c - \mathbf{R}$ by $\mathbf{r}_c - \mathbf{r}_0 - (s - s_0)\mathbf{t}_0$. Note first that by the method used in the proof of equivalence of (6.9) and (6.10), we have

$$\left| \int_{s_0-b}^{s_0+b} \{\mathbf{S}(\mathbf{r}_0 - \mathbf{R}) \cdot [\mathbf{f}(s) - \mathbf{f}(s_0)] \right.$$
$$\left. + \frac{a^2}{4\mu} \mathbf{D}(\mathbf{r}_c - \mathbf{R}) \cdot [\mathbf{f}_n(s) - \mathbf{f}_n(s_0)]\} \, ds \right|$$
$$\leq \frac{K' f_0}{8\pi\mu l} \int_{s_0-b}^{s_0+b} [\,|\mathbf{r}_0 - \mathbf{R}\,|^{-1}$$
$$+ a^2 |\mathbf{r}_c - \mathbf{R}|^{-3}] \, ds \equiv J, \qquad K' = \text{constant} \tag{6.14}$$

Recall now that $\mathbf{r}_c - \mathbf{R} = \mathbf{r}_c - \mathbf{r}_0 - (s - s_0)\mathbf{t}_0 - \frac{1}{2}(s - s_0)^2\mathbf{g}$ and therefore (since $\mathbf{r}_c - \mathbf{r}_0$ is orthogonal to $\mathbf{t}_0$ and $|\mathbf{g}| \leq \sqrt{3}L^{-1}$)

$$|\mathbf{r}_c - \mathbf{R}|^2 \geq a^2 + (1 - \sqrt{3}\epsilon)(s - s_0)^2 - \sqrt{3}L^{-1}(s - s_0)^3 - \tfrac{3}{4}L^2(s - s_0)^4 \quad (6.15)$$

Using (6.15) in (6.14), one finds that

$$J = O\left(\frac{f_0}{\mu}\frac{L}{l}\epsilon^\gamma\right) \quad (6.16)$$

Next, we verify the error made by replacing $\mathbf{r}_c - \mathbf{R}$ by $\mathbf{r}_c - \mathbf{r}_0 - (s - s_0)\mathbf{t}_0$ in (6.12). The mean value theorem is again used to expand $\mathbf{S}(\mathbf{r}_c - \mathbf{R}) - S(\mathbf{r}_c - \mathbf{r}_0 - (s - s_0)\mathbf{t}_0)$ and similarly for $\mathbf{D}$, and an estimate analogous to (6.15) is used. Omitting details, the estimates of the error are found to be $O((f_0/\mu)\epsilon^\gamma)$ and therefore [with (6.16)]

$$H_2 = O\left(\left(1 + \frac{L}{l}\right)\frac{f_0}{\mu}\epsilon^\gamma\right) \quad (6.17)$$

Finally, we must show that

$$\int_{s_0-b}^{s_0+b}\left[\mathbf{S}(\mathbf{r}_c - \mathbf{r}_0 - (s - s_0)\mathbf{t}_0)\cdot\mathbf{f}(s_0) + \frac{a^2}{4\mu}\mathbf{D}(\mathbf{r}_c - \mathbf{r}_0 - (s - s_0)\mathbf{t}_0)\cdot\mathbf{f}_n(s_0)\right]ds$$
$$= \frac{\mathbf{f}_n(s_0)}{4\pi\mu} + \int_{\delta<|s-s_0|<b}\mathbf{S}((s - s_0)\mathbf{t}_0)\cdot\mathbf{f}(s_0)\,ds + O(E) \quad (6.18)$$

By the local analysis (with $c = b$) the left-hand side of (6.18) may be written [see (6.4)]

$$\frac{\mathbf{f}_t(s_0)}{4\pi\mu}\left[\ln\left(\frac{4b^2}{a^2}\right) - 1\right] + \frac{\mathbf{f}_n(s_0)}{8\pi\mu}\left[\ln\left(\frac{4b^2}{a^2}\right) + 1\right] + O\left(\frac{a^2}{b^2}\right)$$
$$= \frac{\mathbf{f}_n(s_0)}{4\pi\mu} + \frac{\mathbf{f}(s_0)}{4\pi\mu}\ln\left(\frac{4b^2}{ea^2}\right) + O\left(\frac{a^2}{b^2}\right)$$
$$= \frac{f_n}{4\pi\mu} + \int_{\delta<|s-s_0|<b}\mathbf{S}((s - s_0)\mathbf{t}_0)\cdot f(s_0)\,ds + O(\epsilon^{2-2\gamma})$$

if $\delta = \frac{1}{2}a\sqrt{e}$.

Thus, we see, combining our estimates, that (6.9) is established with

$$E = \frac{1}{2}\frac{f_0}{\mu}\left(\frac{L}{l} + 1\right)(\epsilon^{\gamma} + \epsilon^{1-\gamma})$$

and taking $\gamma = \frac{1}{2}$ establishes the theorem as stated, this choice implying the smallest error estimate. This completes the proof of the theorem.

6.3 Remarks

Lighthill notes that direct integrations for particular $\mathbf{f}(s)$ suggest that the actual error is $O(\epsilon)$. It is not known if an $\epsilon^{1/2}$ error is realized for an admissible $\mathbf{f}$. [Of course, the main point is the smallness of the error compared to $(-\ln \epsilon)^{-n}$.] Note that if $L = \infty$, the error E becomes indeterminant, whereas if $l = \infty$, the error stays $O(\epsilon^{1/2})$ but the integral on the right of (6.9) diverges logarithmically. For undulating zero-thrust flagella of finite length L_f we can expect the theorem to give a good approximation if the length is rather large compared to a typical wavelength. On the other hand, if the flagellum pushes a large cell body and is therefore delivering nonzero thrust, the resistance of the flagellum as it drifts backward relative to its zero-thrust motion is computed by considering Stokes flow with this drift velocity past the wavy surface. For this calculation the series expansions of Batchelor (1970) or Cox (1970) must be used; these proceed in powers of $\Delta = [\ln (L_f/a)]^{-1}$. Terms (in $\mathbf{u}$) of order Δ and Δ^2 are sufficient to give a series for an inverse resistance coefficient K^{-1} of the form $\alpha\Delta^{-1} + \beta + O(\Delta)$, where α and β depend upon waveform. Thus, the relation between resistance force and translational velocity can be found, but, as we have noted, the error is significantly larger in terms of a/L than that obtained above for the zero-thrust flagellum.

For zero-thrust swimming, (6.9) implies that there is no exactly *local* relation between velocity and resistance; the velocity at a given section depends upon the Stokeslet-dipole distribution (or the force distribution) over the entire flagellum. A semilocal "range of influence" of the force at a given point s is roughly several wavelengths. The range of influence increases with l until essentially the entire flagellum contributes. From these deductions Lighthill concludes that the rational approach to flagellum hydrodynamics must account for the quite different properties of zero-thrust and nonzero-thrust flagella, the latter case involving strongly nonlocal Stokes' flows, the former being at most "semilocal" on the scale of the wavelength.

As a final point it is worth noting that, because we might expect the *velocity* rather than force to be prescribed in applications of (6.9), the problem poses an interesting integral equation for **f**.

6.4 Application to helical waves

This example is significant because of its biological relevance as well as the remarkable fact that the integral equation (6.9) for **f** may be inverted by inspection. Indeed, the force distribution should be invariant under any transformation group that leaves unchanged the equations (5.11) for the helix, in particular the transformation $s \to s + \Delta s$, $x \to x + \alpha\,\Delta s$, $y \to y\cos\theta - z\sin\theta$, $z \to z\cos\theta + y\sin\theta$, $\theta = k\,\Delta s$. This suggests that for zero mean thrust, **f** will have y and z components that are linear combinations of $\sin ks$ and $\cos ks$ whereas the x component, necessarily constant, will vanish. The appropriate phase can be decided by inspection and it is found that the necessary boundary velocity is generated by taking

$$\mathbf{f}(s) = (0,\ h\sin ks,\ -h\cos ks) \tag{6.19}$$

where h is a constant. The velocity of the centerline relative to the fluid at ∞ is **w**, where [see (5.3a) and (5.4)]

$$\begin{aligned}\mathbf{w} &= (U_0 - V)\mathbf{i} + Q\mathbf{t}\\ &= (U_0 - V + Q\alpha,\ -bkQ\sin ks,\ bkQ\cos ks)\\ &= (U_0,\ -bkQ\sin ks,\ bkQ\cos ks)\end{aligned}$$

the subscript on U_0 referring to zero thrust. If the helix rotates with angular speed Ω we have [see Exercise 6.2 and equation (5.11)]

$$\mathbf{w} = (U_0,\ -\omega b\sin ks,\ \omega b\cos ks), \qquad \omega = kQ - \Omega \tag{6.20}$$

Moreover, $\mathbf{f}_n(s) = \mathbf{f} - (\mathbf{f}\cdot\mathbf{t})\mathbf{t} = h\alpha(bk,\ \alpha\sin ks,\ -\alpha\cos ks)$ [recall that $\mathbf{t} = (\alpha,\ -bk\sin ks,\ bk\cos ks)$], and this expression, together with (6.19) and (6.20), may be used to write (6.9) in the form

$$(-U_0,\ wb\sin ks_0,\ -wb\cos ks_0) = \frac{h\alpha}{4\pi\mu}(bk,\ \alpha\sin ks_0,\ -\alpha\cos ks_0)$$
$$+\frac{1}{8\pi\mu}\int_{\rho<\delta}\frac{(0,\ h\sin ks,\ -h\cos ks)\rho^2 - bh\sin k(s - s_0)\rho}{\rho^3}\,ds \tag{6.21}$$

where $\rho = (\alpha(s - s_0), b(\cos ks - \cos ks_0), b(\sin ks - \sin ks_0))$. The x component of (6.21) then gives [with $ks = k(s - s_0) + ks_0 = \theta + ks_0$ in the integrand]

$$U_0 = -\frac{h\alpha bk}{4\pi\mu} + \frac{h\alpha bk}{4\pi\mu}\int_\epsilon^\infty \frac{\theta \sin\theta}{R^3(\theta)}\,d\theta \tag{6.22}$$

where $\epsilon = k\delta$ and $R(\theta) = [\alpha^2\theta^2 + 2(1 - \alpha^2)(1 - \cos\theta)]^{1/2}$. Similarly, the other two components give

$$wb = \frac{h\alpha^2}{4\pi\mu} + \frac{h}{4\pi\mu}\int_\epsilon^\infty \frac{\cos\theta}{R(\theta)}\,d\theta + \frac{h(1-\alpha^2)}{4\pi\mu}\int_\epsilon^\infty \frac{\sin^2\theta}{R^3(\theta)}\,d\theta \tag{6.23}$$

Thus, (6.9) can be satisfied for these values of U, w and the assumed form (6.19) of $\mathbf{f}$ is verified. Evaluating these integrals for small ϵ, there results

$$U_0 = \frac{h\alpha bk}{4\pi\mu}[-1 - \log\epsilon + A_1(\alpha) + O(\epsilon)]$$

$$wb = \frac{h}{4\pi\mu}[\alpha^2 - 1 - (2 - \alpha^2)\log\epsilon + \alpha^2 A_1(\alpha) \tag{6.24}$$

$$+ 2(1 - \alpha^2)A_2(\alpha) + O(\epsilon)] \tag{6.25}$$

The reader may refer to Lighthill's paper for graphs of the functions A_1 and A_2.

6.5 Comparison with resistive-force theory

The tangential component of velocity is $(-U_0, +bhQ\sin ks, -bhQ\cos ks)\cdot(\alpha, -bk\sin ks, bk\cos ks) = -\alpha U_0 - \omega b^2 k$, whereas the tangential *force* is $-bkh$. Thus, by definition $K_T =$

Figure 6.1. A graph allowing comparison of several approximate theories of zero-mean thrust swimming of a helix (see text).

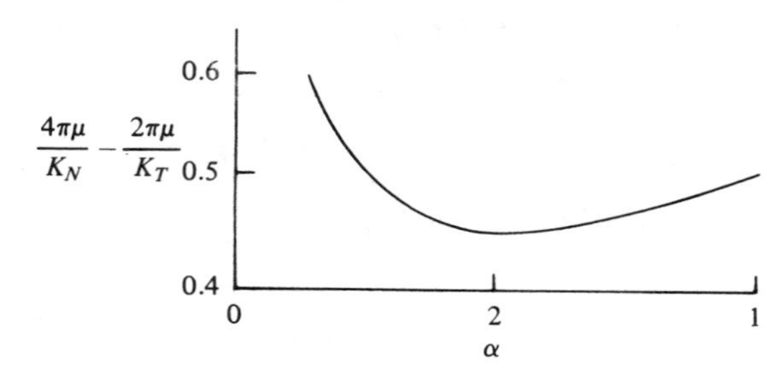

$bkh(\alpha U_0 + \omega b^2 k)^{-1}$, and similarly $K_N = h\alpha(\omega b\alpha - U_0 bk)^{-1}$.From (6.24) and (6.25) we may then evaluate $4\pi\mu/K_N - 2\pi\mu/K_T = \frac{1}{2} + (1 - \alpha^2)[A_2(\alpha) - A_1(\alpha)]$. This quantity would be 1 (for any bc) if (6.5) applied and would be $\frac{1}{2}$ (for any q) if (6.6) applied. The exact values, given by the foregoing function of α, are shown in Figure 6.1. It is seen that the value of $\frac{1}{2}$ of Shack et al. (1974) is a good constant approximation, whereas the Gray–Hancock value of 1 is not. For additional discussion of this point and a penetrating interpretation of the discrepancy of the Gray–Hancock value, see Lighthill (1979).

Exercises

6.1 Show that the mean torque per unit flagellum length exerted by the force (6.19) is $-bh\mathbf{i}$ and that the rate of working in zero-thrust swimming is accordingly ωbh.

6.2 Verify (6.17).

7

Ciliary propulsion

7.1 Introduction

This is the second basic swimming mechanism in the Stokesian realm. Although the organelles are apparently identical in ultrastructure and physiology, we use the term "flagellum" when there is only one, or a small number of these hairlike appendages on a cell, as in spermatozoans and the flagellates, and "cilia" to denote large numbers of them on the same cell, as in the ciliates. Cilia, then, are just a large number of flagella on the same cell. The ciliates tend to be larger than the cell bodies of spermatozoans by an order of magnitude. Cilia tend to be shorter than flagella, however, and this fact puts their mechanism firmly in the Stokesian realm.

Confusion about terminology can arise from two sources: (1) in bacteria there are organelles called flagella with very different ultrastructure and physiology (Chapter 4); and (2) the term "cilia" is usually used in connection with various ciliated tissues in metazoans (many-celled animals) as in the lining of our respiratory tracts, but there is usually only one or a small number of cilia per cell in these tissues.

Because of their proximity to the cell wall, we can expect to find that optimal propulsion using cilia will involve movements quite different from those of a single large flagellum. In this regard it would be valuable to understand the possible adaptive advantages that could have led to the evolution of ciliary motion. As both flagellary and ciliary modes are widespread and both are evidently successful strategies, we might suppose that there are trade-offs between them.

In one sense the proliferation of hairlike appendages is an efficient use of material for the development of large resistance. A square plate of side L and thickness h will experience a resistance in Stokes' flow that is some multiple of μUL. However, if the plate is broken into N equal "hairs" of length L (Figure 7.1), the force experienced by each hair (Chapter 5) is roughly $\mu UL/\ln(L/h)$; the total force is $\mu UL^2/h \ln(L/h) \gg U\mu L$ if $L/h \gg 1$. Thus, a Stokesian parachute to be constructed of a fixed amount of material should consist of thin rigid filaments. Such a strategy is probably seen in seed dispersal by plants such as dandelions and in the membraneless wings of certain minute flying insects.

Such considerations alone do not, however, suffice to explain the possible value of ciliary propulsion to the organism. As we have seen in Chapter 5, the mechanical efficiency of a flagellum depends on resistances largely through their *ratio* ρ (according to Gray–Hancock theory), and the optimal efficiencies for isolated flagella tend to be 1 to 10 percent, with 4 percent typical. Because of the prevalence of both strategies, it is perhaps not surprising that estimates of efficiency for ciliary propulsion tend to be comparable [see, e.g., Blake (1971*b*, 1974) and Exercise 7.3] Although such estimates are crude, the fact is that efficiencies are rather unexpectedly high in view of the adverse hydrodynamic interaction between cilium and cell wall. Apparently, these losses can be compensated by optimizing the movement of the cilium at a given location on the cell wall.

This last point suggests that the most compelling advantage of a ciliary system is its flexibility. A large number of essentially identical organelles can be distributed and coordinated as required for optimal propulsion of a cell body of given shape and size. In addition, cilia may play an important role in generating currents nearer an oral cavity for feeding, a role similar to that of the ciliated tissues in metazoans; we shall not explicitly pursue this latter topic, however.

Possible disadvantages of the ciliary mode include the clear need to coordinate the motions of a large number of organelles.

7.1.1 *Motion of a cilium*

Typical parameters for a single cilium are: length $\sim$ 0.001 cm = 10 μm (1 μm = 10^{-4} cm); diameter $\sim$ 0.25 μm; tip speed $\sim$ 0.2 cm/sec; beat frequency $\sim$ 30 cycles/sec. In a typical motion a

Figure 7.1. Square plate broken into equal-length "hairs."

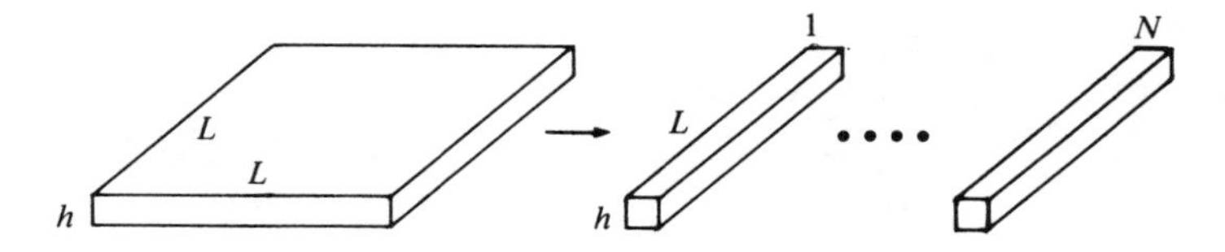

ciliary beat has two parts, an *effective* stroke and a *recovery* stroke (Figure 7.2). The effective stroke may take somewhat less than one-half of a cycle, and a Reynolds number for a single cilium is length $\times$ tip speed/viscosity $\sim$ 0.02. The Reynolds number based on ciliary diameter is smaller by a factor of $\frac{1}{40}$.

Cilia may beat in a vertical plane, but in some cases the recovery stroke involves motions sideways motions to the cell surface. In some species cilia may move in a bundle, comprising a *compound cilium*. Helical waves are also observed.

The basic motion seems sound from a fluid mechanical viewpoint. The effective stroke moves the cilium broadside or, pivoting about its base, developing the larger force associated with the coefficient of normal resistance K_N. During recovery the motion is mainly tangential and the resistance is smaller by a factor of roughly $\frac{1}{2}$. However, many variations of the basic stroke are seen and the foregoing remarks have neglected hydrodynamic interaction with the cell wall [see e.g., Jahn and Votta (1972)].

7.1.2 *Metachronal coordination*

We shall now summarize the observed organization of beating of a large array of cilia. This *metachronal* coordination may take several forms. Figure 7.3*a* shows *symplectic* metachrony, in which the effective stroke moves in the same direction as wave crests of the cilia tips. The opposite situation, *antiplectic* metachrony, occurs when the effective stroke and crest velocity are in opposite directions (Figure 7.3*b*). It is generally observed that symplectic metachrony involves less horizontal movement of the cilia tips than does antiplectic metachrony. Additional complications arise from the possibility that the phase of cilia may vary in a lateral direction, so that waves can propagate obliquely or even at right angles to the plane of the beat. The terminology is summarized in Figure 7.4.

Figure 7.2. Stroke configurations. (*a*) Effective. (*b*) Recovery.

A typical protozoan with symplectic metachrony is the species *Opalina ranarum*, a disc-shaped organism inhabiting the gut of a frog. It is about 200 μm thick and the cilia are 10 μm in length, arranged in rows 3 μm apart. Within each row the distance between cilia is about 0.3 μm. Beat frequency is only 5 cycles/sec. The metachronal wavelength is 30 to 50 μm and the metachronal wave speed (speed of crests) is 100 to 200 μm/sec. The cell swimming speed is about 50 μm/sec, about one-fourth of the wave speed.

Typical dexioplectic metachrony is seen in *Paramecium* (which, together with *Opalina*, has been widely studied in connection with cil-

Figure 7.3. Metachrony configurations. (*a*) Symplectic: side view, planar motion. (*b*) Antiplectic.

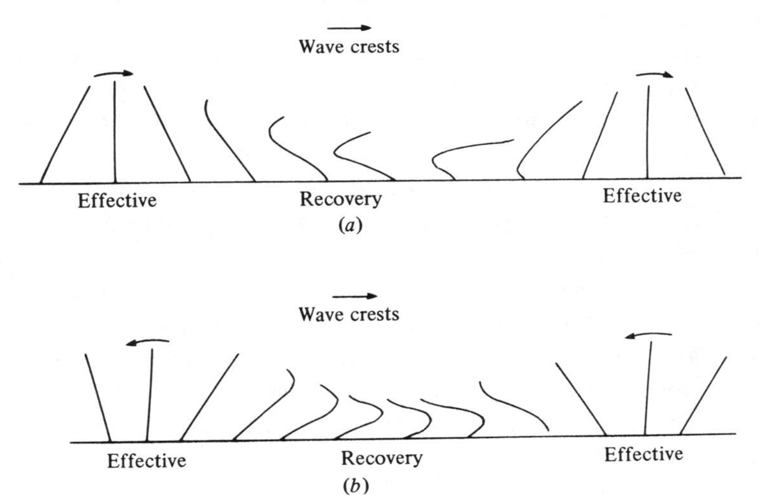

Figure 7.4. Terminology for an array of cilia, top view. (After Blake and Sleigh, 1975.)

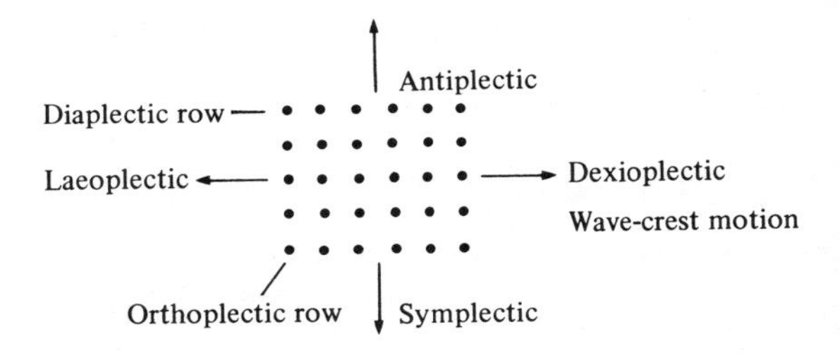

iary propulsion). This cigar-shaped protozoan, which is roughly 200 μm long, has cilia of length 12 μm arranged in a square array about 2 to 3 μm apart, beating with a frequency 20 to 30 cycles/sec. Metachronal wavelength is only 10 μm, the wave speed being about 200 μm/sec. Whereas the last figure is comparable to *Opalina,* the swimming speed of *Paramecium* is 1000 to 2000 μm/sec, 5 to 10 times the wave speed. Models for ciliary propulsion must account for this wide range of behavior among ciliates.

7.2 The envelope model

This model takes the view that a closely spaced array of waving cilia, beating at a sufficiently small Reynolds number, will move the nearby fluid very much as if the ciliary tips were replaced by their continuous envelope, to which the adjacent fluid adheres. Putting aside for the moment the validity of this picture, the model clearly realizes the continuous extensible, undulating sheet studied in Chapter 3; we therefore base our analysis on the results given there. The motion of individual cilia plays no direct role and the parameters that enter are those of the *metachronal waves.* The problem we treat assumes, as in Chapter 3, that the envelope is infinite and the undulations two-dimensional.

The model assumes further that the envelope is impermeable to the fluid, although of course this is not precisely true. We also should note particularly that the assumptions are more likely to hold for symplectic metachrony, in which the ciliary tips remain close together, than for antiplectic metachrony, in which the tips are isolated during the effective stroke (see Figure 3.3).

We shall use the main results of Chapter 3 for a Stokesian sheet with coordinates satisfying

$$x_s = x + a\cos(kx - \omega t - \phi)$$
$$y_s = b\sin(kx - \omega t) \tag{7.1}$$

Taking a typical point by setting $x = 0$, we see that

$$(bx_s - a\sin\phi y_s)^2 + a^2\cos^2\phi y_s^2 = a^2b^2\cos^2\phi \tag{7.2}$$

The orbit is thus an ellipse. Setting $x_s = aX$, $y_s = bY$, we get

$$X^2 - 2XY\sin\phi + Y^2 = \cos^2\phi \tag{7.3}$$

The form of the locus for various ϕ, b/a can therefore be found from (7.2) or (7.3), and we give some examples in Figure 7.5.

Let us now relate the parameters of the envelope to the metachronal wave. Both ω and k are metachronal wave parameters. (Since $k > 0$ in the analysis of Chapter 3, we adopt that convention here.) Then from (7.1) the metachronal waves move to the right of $\omega > 0$. From Figure 7.5 we see that with $\phi = 0$, the "cilia tip" moves to the right when farthest from the sheet and therefore corresponds to symplectic metachronism; the case $b > a$ is perhaps preferred because, as noted above, this metachronism typically has more vertical than horizontal tip movement. Similarly, $\phi = \pi$, $b < a$ would give a typical case of antiplectic metachronism.

From the expressions for swimming speed and swimming effort,

$$U = \tfrac{1}{2}\omega k(b^2 + 2ab\cos\phi - a^2) \tag{7.4}$$

$$w_s = \mu\omega^2 k(a^2 + b^2) \tag{7.5}$$

we note that $U > 0$ if $\phi = 0$ and $b > (\sqrt{2} - 1)a$. That is, the sheet swims to the left, opposite the direction of propagation of the metachronal wave, if cilia tips move mainly up and down, again in rough accord with observations of organisms such as *Opalina*. Similarly, if $\phi = \pi$ and $a > (\sqrt{2} - 1)b$, the sheet moves to the right, opposite the direction of propagation of the metachronal wave, in accord with observations of organisms such as *Paramecium*.

Figure 7.5. Tip loci in the envelope model; the direction of the orbit is for ω, a, $b > 0$.

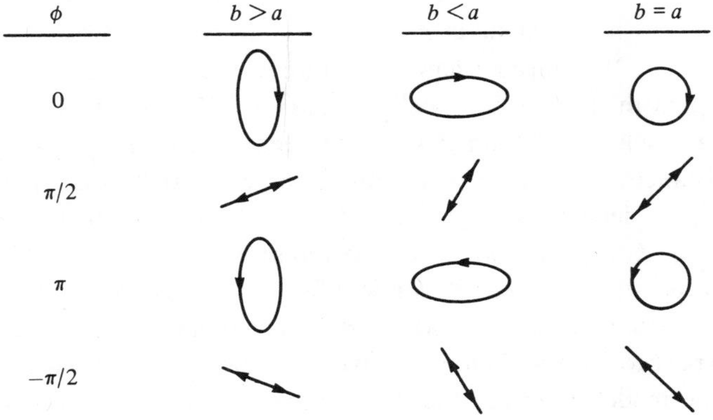

Taken together, (7.4) and (7.5) pose an interesting optimization problem: Minimize the work (per unit area) required for a given swimming speed U, or equivalently maximize $|U|$ for fixed W_s. We suppose that ω and k are fixed, and set $b = R \cos \theta$, $a = R \sin \theta$, so that R is also fixed. We then seek extremals of $\cos^2 \theta + 2 \cos \theta \sin \theta \cos \phi - \sin^2 \theta = \cos 2\theta + \cos \phi \sin 2\theta$ with respect to variations of θ and ϕ. Differentiating, we get $-\sin \phi \sin 2\theta = 0$, so that either $\phi = 0$ or π, or $\sin 2\theta = 0$. The derivative with respect to θ then gives $-2 \sin 2\theta + 2 \cos \phi \cos 2\theta = 0$, so that either $\sin 2\theta = 0$ *and* $\cos \phi = 0$ or else one of the two cases $\phi = 0, 2\theta = 1$ or $\phi = \pi, 2\theta = -1$ will occur. The first choice implies that either $\sin \theta$ or $\cos \theta$ vanishes, so either a or b is zero. The second choice ($\phi = 0$) gives $2 \tan \theta = 1 - \tan^2/\theta$ or $\tan \theta = -1 \pm \sqrt{2}$, so either $a(\sqrt{2} - 1)b$ or $b = (1 - \sqrt{2})a$. The last choice ($\theta = \pi$) gives $2 \tan \theta = \tan^2 \theta - 1$, so either $a = (1 - \sqrt{2})b$ or $b = (\sqrt{2} - 1)a$ As we may assume that $a, b > 0$ without loss of generality, we obtain swimming speeds $\pm\frac{1}{2} \omega k R^2$ if a or b vanishes and

$$U = \frac{1}{\sqrt{2}} \omega k R^2, \qquad \phi = 0, \quad a = (\sqrt{2} - 1)b \qquad \text{(simplectic)}$$

$$U = -\frac{1}{\sqrt{2}} \omega k R^2, \qquad \phi = \pi, \quad b = (\sqrt{2} - 1)a \qquad \text{(antiplectic)}$$

Thus, a speed gain of about 41 percent results from a combination of vertical and horizontal movements of the envelope, and the optimal parameters fall into the range of those already noted to be consistent with observation.

7.3 Macrostructure

We consider now the nature of the flow field about a ciliated organism. Define L = cell diameter $\sim$ 200 μm, λ = metachronal wavelength $\sim$ 40 μm, l = cilium length $\sim$ 10 μm, d = interciliary distance $\sim$ 3 μm, c = cilium diameter $\sim$ 0.25 μm. These lengths form a decreasing sequence and we shall be concerned here with the scales L, λ, and l, which we take to satisfy $L \gg \lambda \gg l$ (Figure 7.6). Recall that the hydrodynamic effects of the swimming sheet extend outward a distance $O(\lambda)$ (with exponential decay). We therefore expect to have a "boundary layer" of thickness $\sim \lambda$ where the envelope model describes the flow, provided that the parameters of the envelope are given a certain dependence on position on the scale L of the cell body. The parameters would presumably be determined from the analysis of a *sublayer* of thickness $O(l)$ adjacent to the cell body (see below). For our present purposes, we can regard the envelope as hydrodynamically equivalent to the sublayer insofar as the macrostructure associated with scales λ and L is concerned. This viewpoint is the basis for a theory of swimming of ciliates developed by Brennan (1974, 1975).

From the local analysis of the slowly varying sheet given in Chapter 3, we can now piece together the flow field on the scale of the cell body as follows. Relative to an observer at rest with respect to the fluid at infinity, the velocity field $\mathbf{U}$ satisfies ($\mathbf{U}_t$ and $\mathbf{U}_n$ are tangential and normal vectors)

$$\mathbf{U}_n|_s = O\left(\frac{\lambda}{L}\right) \simeq 0 \tag{7.6}$$

$$\mathbf{U}_t|_s = \mathbf{U}_0(S) + \mathbf{U}_s + \mathbf{r}_s \times \mathbf{\Omega} \tag{7.7}$$

where $\mathbf{U}_0(s)$ is the tangential vector $k^{-1}U_0\mathbf{k}$, $\mathbf{k}$ being the metachronal wave vector and U_0 the function given [for waves (6.1)] by (6.4). Here $\mathbf{k}$, ω, a, b, and ϕ are arbitrary functions of position. The vectors $\mathbf{U}_s$ and $\mathbf{\Omega}$ are the instantaneous velocity and angular velocity with respect to the centroid of the cell body.

Also, in the force and torque balance, the "envelope stresses" σ_t and σ_n (derived at the end of Chapter 3 for flow in two dimensions) must be included. The resulting equations determine U_s and $\mathbf{\Omega}$ at each instant. To apply this approach to three-dimensional bodies we must, however, extend the expressions for σ_t and σ_n to an arbitrary surface. Let us assume that the metachronal wave is fully determined by a

Figure 7.6. Macrostructure of a ciliate, $L \gg \lambda \gg l$.

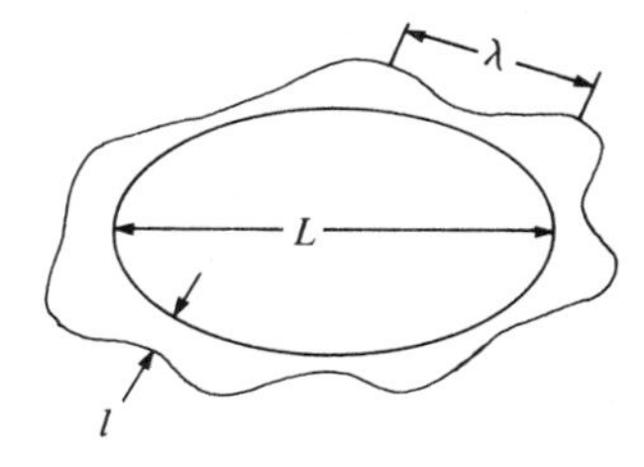

slowly varying wavenumber vector $\mathbf{k}$ and slowly varying parameters ω, b, γ, and β; then (3.26*a*) and (3.26*b*) provide the local expressions

$$\iint_{\Delta S} \sigma_t \, dS = \int_c \alpha_t \mathbf{k} \cdot \mathbf{t}_1 \, ds, \qquad \alpha_t = -2\mu\omega\gamma b \tag{7.8}$$

$$\iint_{\Delta S} \sigma_n \, dS = \int_c \alpha_n \mathbf{k} \cdot \mathbf{t}_1 \, ds, \qquad \alpha_n = \mu\omega\beta b \tag{7.9}$$

when applied to a small area element ΔS bounded by the closed curve C, $\mathbf{t}_1$ being a vector in the tangent plane that is also an outer normal to C. Now let $\mathbf{t}_2$ be another vector in the tangent plane, orthogonal to $\mathbf{t}_1$ and pointing in the direction of increasing arc length s, and let $\mathbf{n}$ be the normal to ΔS coinciding with $\mathbf{t}_1 \times \mathbf{t}_2$ on C. Then we have

$$\int_C \alpha \mathbf{k} \cdot \mathbf{t}_1 \, ds = \int_C \alpha(\mathbf{n} \times \mathbf{k}) \cdot \mathbf{t}_2 \, ds = \iint_{\Delta S} \mathbf{n} \cdot [\nabla \times \alpha \mathbf{n} \times \mathbf{k}] \, dS$$

by Stokes' theorem, and therefore

$$\sigma_{t,n} = \mathbf{n} \cdot [\nabla \times \alpha_{t,n}(\mathbf{n} \times \mathbf{k})] \tag{7.10}$$

7.3.1 *The ciliated sphere*

To carry out this construction in a particular case, we consider the axisymmetric ciliated sphere shown in Figure 7.7. We suppose that ω, k, and ϕ are constant; that the crests of the metachronal waves are lines of constant latitude; and that $a = a_0(\sin\theta)^{1/2}$, $b = b_0(\sin\theta)^{1/2}$. The net force exerted on the sphere by the "envelope stress" is

$$-F = 2\pi r_0^2 \int_0^\pi (\sin^2\theta \, \sigma_t - \sin\theta \cos\theta \, \sigma_n) \, d\theta$$

Figure 7.7. Ciliated sphere.

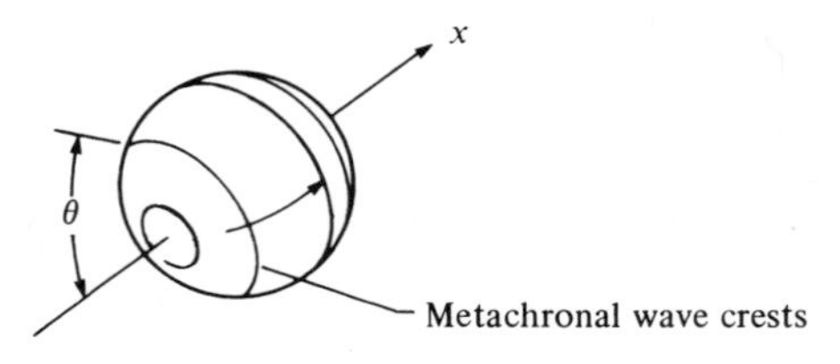

where r_0 is the radius of the sphere. Applying (7.10) to the sphere, we have

$$\sigma_t = -2\mu\omega k a_0 b_0 \sin\phi \left(\frac{1}{r_0 \sin\theta} \frac{\partial}{\partial\theta} \sin^2\theta \right)$$

$$= -4\mu\omega k a_0 b_0 \sin\phi \cos\theta \; r_0^{-1}$$

$$\sigma_n = 2\mu\omega k a_0 b_0 \cos\phi \cos\theta \; r_0^{-1}$$

Consequently,

$$F = -\tfrac{8}{3}\pi\mu\omega k r_0 a_0 b_0 \tag{7.11}$$

Thus, in steady motion the Stokes flow relative to the fluid at infinity due to the action of the cilia and the envelope stress will contain a Stokeslet $F\mathbf{S} \cdot \mathbf{i}$, representing a force balancing (7.11) *applied to the fluid*. The velocity field then has the form

$$\mathbf{u} = \frac{F}{8\pi\mu}\left(\frac{\mathbf{i}}{r} - \frac{\mathbf{r}\cos\theta}{r^2}\right) + r_0^3 A \nabla\left(\frac{\cos\theta}{r^2}\right) \tag{7.12}$$

The boundary conditions on $r = r_0$ are

$$u_\theta = (U_s + U_0)\sin\theta \tag{7.13a}$$

$$u_r = -U_s \cos\theta \tag{7.13b}$$

where U_s is the (constant) swimming velocity and

$$U_0 = \tfrac{1}{2}\omega k(b_0^2 + 2a_0 b_0 \cos\theta - a_0^2)$$

From (7.12) and (7.13) we get

$$\frac{F}{8\pi\mu r_0} - A = U_s + U_0, \qquad \frac{F}{4\pi\mu r_0} + 2A = U_s$$

Eliminating A, we obtain the swimming velocity,

$$U_s = -\tfrac{2}{3}U_0 + \frac{F}{6\pi\mu r_0} = -\tfrac{1}{3}\omega k(b_0^2 + \tfrac{10}{3}\, a_0 b_0 \cos\phi - a_0^2) \tag{7.14}$$

Note that if $a_0 = 0$, the sphere swims at two-thirds the speed of an infinite sheet with these parameters. We also find that $|U_s|/\omega k(a_0^2 + b_0^2)$ has a maximum value of about 0.648 compared to 0.707 for the infinite sheet. A number of such examples have been worked out by Brennen (1974), who also compares the predicted swimming speeds with observations.

As a final point we may compare the ratio of maximum swimming speed to phase speed, equal to $0.648k^2(a_0^2 + b_0^2)$, with observation. If we take $a_0^2 + b_0^2$ to equal one-fourth of the square of the cilium length, for *Opalina* the ratio is $(2\pi)^2(0.65/4)(10^2/40^2) = 0.40$. The observations indicate that $k|U_s|/\omega \simeq 0.25$, so the theory is reasonably close. For *Paramecium* these estimates are off by an order of magnitude, and it is generally believed that the failure of the envelope model in that case is really due to the lack of a direct knowledge of the envelope's parameters, presumably because of the wide tip separation of the ciliary tips during the effective stroke. Intuitively, the wide tip separation would suggest that in antiplectic metachronism the interaction of cilia would be less pronounced, and this could account for the higher swimming speeds. In any event, the postulate of an impermeable surface to which the fluid adheres, essentially equivalent to the ciliary tip envelope, would be untenable. With these points in mind, we turn to the structure of the flow field on the level of the cilium length.

7.4 Microstructure: the sublayer

The basic goal is a direct attack on the problem of an array of beating cilia attached to a plane wall. A full analysis of this problem should, among other things, provide the equivalent "envelope," representing the flow field on the scale of the metachronal wavelength given the appropriate large-scale organization of the cilia. In practice the problem has been formulated in such a way that the outer flow is "contained" in the solution for the microstructure. The swimming speed is then obtained by evaluating the fluid speed well above the array.

A model for the microstructure, called the *sublayer* model, has been devised by Blake (1972). A related *traction layer* model has been put forward by Keller et al. (1975). Both of these models attempt to deduce the mean velocity profile as a function of vertical distance through the ciliated layer, from the prescribed motion of typical single cilium.

We shall be concerned here only with the sublayer model. In the sublayer model the cilium is represented by appropriate distribution of singular solutions (Stokelets and source dipoles, as discussed in Chapter 6) but now supplemented with an image system representing the effect of the wall [see, e.g., Lighthill (1975, p. 129)]. In this way one computes a Green's tensor $\mathbf{G}(\mathbf{r}, \mathbf{R})$. That is, $\mathbf{F} \cdot \mathbf{G}$ is the velocity field at $\mathbf{r}$ due to force $\mathbf{F}$ localized at $\mathbf{R}$, in the presence of a plane wall to which the fluid adheres. Summing over an infinite (rectangular) array of cilia, we have

$$\mathbf{u} = \sum_{k=1}^{\infty} \int_0^L \mathbf{F}_k(\mathbf{R}_k(s, t)) \cdot \mathbf{G}(\mathbf{r}, \mathbf{R}_k(s, t)) \, ds \tag{7.15}$$

Now $\mathbf{F}_k$ may be related to the local incident velocity by the Gray–Hancock theory. On the other hand, the velocity incident onto a given cilium is produced by the combined effect of the array. An integral equation is implied, and it is natural to introduce an average over horizontal coordinates and time. Of course, as is customary in problems of this kind, the time average does not lead to a closed system, because the difference between exact and time mean values (i.e., the fluctuating part of any variable) must itself be expressed in terms of mean variables.

If the cilia move in such a way that they may be determined by their vertical coordinate $Z(s, t)$, the averaging yields

$$\begin{aligned} \left\langle \mathbf{u} \right\rangle &= \mathbf{U}(z) \\ &= (\Delta x \, \Delta y)^{-1} \left\langle \int_0^L K(z, Z(s, t)) \mathbf{F}(\mathbf{R}(s, t)) \, ds \right\rangle \end{aligned} \tag{7.16}$$

with an error $O(\Delta x \, \Delta y \omega / l)$, where Δx and Δy are sides of a rectangle of the array. The kernel $K(z, Z)$ represents the effect of a concentrated constant horizontal force of unit magnitude per unit area at the plane $z = Z$, in the presence of a rigid wall at $a = 0$. We have, with K continuous,

$$\mu \frac{d^2 K}{dz^2} = \delta(z), \qquad K(0, Z) = 0, \quad K(\infty, Z) \text{ bounded}$$

Therefore,

$$K = \begin{cases} \mu^{-1} z & \text{if } z < Z \\ \mu^{-1} Z & \text{if } z > Z \end{cases}$$

The principal difficulty of the method comes from the form taken by the incident velocity onto a point of a cilium. This breaks up naturally into the global mean velocity $\mathbf{U}$ plus a term representing the local influence of the given cilium on the local mean velocity (these two terms are those of a single cilium beating in the mean flow **U**), plus the fluctuating component. If the fluctuating component is neglected, an assumption that Blake makes, the incident velocity is essentially an expectation conditioned by the presence of a single cilium, and the theory, in relating the force in (7.16) to the incident velocity, then closely parallels Gray–Hancock theory. This procedure leads to two integral equations for **U** and the local velocity correction **V** for a single cilium. The input for numerical computations is then the metachronal parameters and the beat pattern of the cilium.

Blake applies his theory to symplectic and antiplectic metachronism, but in these cases he modifies (7.16) by a weight function in the integrand. However, recently Liron and Mochon (1976) have argued that (7.16) is correct as it stands, and have taken a somewhat different approach to solve for **F** (see below). The errors associated with the omission of the fluctuating component have been discussed by Keller et al., (1975). They replace the discrete array of cilia by a continuous one, thus allowing the nonstationary components to be followed in detail.

Liron and Mochon make the interesting observation that if horizontal averaging is relaxed in favor of averaging in the direction orthogonal to the plane of the beat, then the average fluid velocity in the sublayer (now depending upon x, z, t) coincides at points of the cilium with *its* instantaneous velocity. The partially averaged system they use would therefore appear to describe, in effect, a linear array of waving sheets. The resulting kernel is more complicated than K, because it involves a summation over the linear array of the cilia with differing phase, but the form of the integral equation fully encompasses fluctuating components.

The sublayer models are successful in explaining many of the features of ciliary propulsion. Applied to *Opalina,* Blake's calculations give a swimming speed (of the array) of about $0.2|k|l$ times the phase speed. With $|k|l$ typically $\pi/2$, this gives the observed ratio of about $\frac{1}{4}$. For *Paramecium* $l/x \sim \frac{6}{5}$ and U is found to be about an order of magnitude greater than the phase speed, thus confirming the role of antiplectic or dexioplectic metachronism in rapid swimming. It is interesting that the models predict, in certain cases, reversed flow near the cell wall. Some recent observations supporting this result have been reported by Cheung and Winet (1975).

Exercises

7.1 Give the entries in Figure 7.6 corresponding to $\phi = \pi/6$.

7.2 Define an efficiency of the envelope model by

$$\eta = \frac{\mu U^2 k}{W_s}$$

Show that η may be twice the value for vertical motion ($a = 0$) for a certain combination of vertical and horizontal tip movement.

7.3 Define an efficiency of the ciliated sphere by

$$\eta = \frac{6\pi\mu r_0 U_s^2}{W_s}$$

Where W_s is computed by integrating (7.5) over the sphere. Show that for $a^2 + b^2 = (a_0^2 + b_0^2)\ \sin\theta$, the optimal efficiency is about

$$5\left(\frac{a_0^2 + b_0^2}{r_0\lambda}\right), \qquad \lambda = \frac{2\pi}{k}$$

For *Opalina* we might take $r_0 = 80\ \mu\text{m}$, $\lambda = 40\ \mu\text{m}$, and $a_0^2 + b_0^2 = \frac{1}{4}(10)^2\ \mu\text{m}^2$, in which case $\eta_{max} \simeq 4$ percent.

8

The Eulerian realm: the inertial force

8.1 The inviscid limit

The second major subdivision of hydrodynamics is the theory of large-Reynolds-number flows. Returning to the dimensional form of the Navier–Stokes equations,

$$\frac{\partial \mathbf{u}}{\partial t} + \mathbf{u} \cdot \nabla \mathbf{u} + \frac{1}{\rho} \nabla p - \nu \nabla^2 \mathbf{u} = \mathbf{g}, \qquad \nabla \cdot \mathbf{u} = 0 \tag{8.1}$$

we let $\nu \to 0$ (in dimensionless theory, $\mathrm{Re} \to \infty$). This limit process is involved. If we simply put $\nu = 0$, the viscous stress term disappears from the dynamical description of the fluid. But if the resulting $\mathbf{u}$ is found to have second derivatives that are unbounded at some point, this may not be the appropriate limit, because $\nu \nabla^2 \mathbf{u}$ need not be small locally irrespective of ν. Such singular behavior necessarily occurs near the boundary of a swimmer. Indeed, $\nu = 0$ implies that the fluid has no tendency to adhere to a surface and slips freely along it, whereas in reality there are necessarily fluid particles adjacent to a boundary that adhere. We must therefore admit a boundary layer theory that complements the inviscid or perfect fluid limit $\nu = 0$. Technically, these difficulties can be traced to the fact that $\nu \to 0$ is a singular limit of a parabolic system of partial differential equations. A second-degree vector equation is reduced to first degree, in either case supplemented by the incompressibility condition.

As the relevant boundary-layer theory would appear to be an inescapable complement to high-Reynolds-number hydrodynamics, the plan to be followed here, namely to build a theory without explicit regard for the viscosity of the fluid, seems highly suspect. In fact, the idealized problem $\nu = 0$ has a certain basic validity for problems of animal locomotion at high Reynolds number, usually because of the nonstationary character of the problem. In any case, the idealized problems to be considered below provide a basis for studying real fluid flows. It turns out that certain aspects of real fluid behavior can be realized by appropriate models with $\nu = 0$ (compare the Kutta condition of Chapter 9). This is the case with the boundary conditions for the inviscid theory.

Setting $\nu = 0$ in (8.1), we have *Euler's equations* for an incompressible, inviscid fluid:

$$\frac{\partial \mathbf{u}}{\partial t} + \mathbf{u} \cdot \nabla \mathbf{u} + \frac{1}{\rho} \nabla p = \mathbf{g}, \qquad \nabla \cdot \mathbf{u} = 0 \tag{8.2}$$

Since ρ and $\mathbf{g}$ are constant, both may be absorbed into a pressure and there are no further principal parameters. Thus, the time derivative of $\mathbf{u}$ and the nonlinear term $\mathbf{u} \cdot \nabla \mathbf{u}$ are inescapable ingredients in the dynamics. (Sometimes geometrical considerations, such as thinness, can be used to effectively suppress nonlinear terms; see, for example, the slender-body theory of Chapter 10.)

Propulsive mechanisms based upon the perfect fluid equations (8.2) depend essentially upon the *reaction* of the fluid to acceleration (as might be experienced by an undulating body). This principal basis for locomotion in the Eulerian realm is thus totally different from the *resistive* forces of the Stokesian realm, the latter being the key concept in flagellar motion, for example. The terms

$$\frac{\partial \mathbf{u}}{\partial t} + \mathbf{u} \cdot \nabla \mathbf{u}$$

equal to the acceleration of a fluid particle per unit volume, fully describe this reaction. We shall adopt a rather unconventional terminology, however. If the vector identity

$$\mathbf{u} \cdot \nabla \mathbf{u} = \nabla \left(\tfrac{1}{2} u^2\right) - \mathbf{u} \times \nabla \times \mathbf{u} \tag{8.3}$$

is used in (8.3), we may write

$$\rho \frac{d\mathbf{u}}{dt} = \underbrace{\rho \left(\frac{\partial \mathbf{u}}{\partial t} + \nabla \left(\frac{1}{2} u^2 \right) \right)}_{\text{inertial force}} - \underbrace{\rho \mathbf{u} \times \omega}_{\text{vortex force}} \tag{8.4}$$

where

$$\omega = \nabla \times \mathbf{u} = \text{fluid vorticity} \tag{8.5}$$

The principal concern of this chapter will be the role of the inertial force as defined in (8.4). Strictly speaking, the vortex force also arises from fluid inertia, but the properties of the vorticity field make this

Figure 8.1. Illustration of Bernoulli's theorem. Gas flows steadily through a contraction, so the velocity increases. As H [equation (8.6)] is constant, pressure decreases across the contraction.

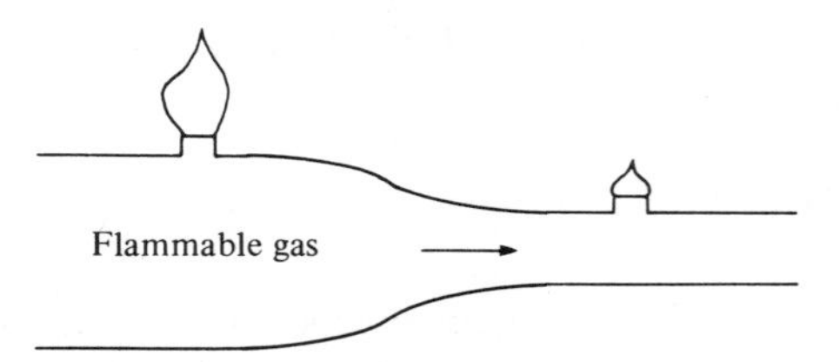

splitting a useful one. (The vortex force will be studied in Chapter 9†.) One immediate consequence of (8.4) follows when used in the dot product of $\mathbf{u}$ and (8.2):

$$\frac{\partial}{\partial t}\left(\frac{1}{2}u^2\right) + \mathbf{u}\cdot\nabla H = 0, \qquad H = \frac{p}{\rho} + \frac{1}{2}u^2 \tag{8.6}$$

For steady (time-independent) motion, this gives $\mathbf{u}\cdot\nabla H = 0$, which tells us that the quantity H is constant along the *streamlines* of the flow. This property (of the inertial force) is known as *Bernoulli's theorem;* H is often referred to as Bernoulli's function. Bernoulli's theorem says that in steady flow the pressure at a streamline varies as minus the square of the speed. The classic experiment is sketched in Figure 8.1.

8.1.1 *Boundary conditions*

At a moving solid boundary, the absence of viscosity relaxes the requirement of adherence of the fluid. Nevertheless, if a boundary is impenetrable (as we shall always assume to be the case here), a fluid particle on the boundary moves tangentially. If

$$S(x, y, z, t) = 0 \tag{8.7}$$

is the implicit equation for the boundary surface S, (8.7) must be invariant on the trajectory of a fluid particle, so that

$$\frac{dS}{dt} = \frac{\partial S}{\partial t} + \mathbf{u}\cdot\nabla S = 0 \tag{8.8}$$

Note that for stationary surfaces, (8.8) is equivalent to the vanishing of the normal component of $\mathbf{u}$ on S.

†The division is rather artificial, however, in flows where the vorticity field is time dependent (see Chapter 11).

8.2 Potential flow

To conveniently isolate the inertial force as defined in (8.4), we consider the important class of *irrotational* flows (zero vorticity). From (8.5) there exists a *potential* ϕ such that $\mathbf{u} = \nabla\phi$. The solenoidal property then implies that $\nabla^2\phi = 0$. With (8.4) in (8.2) and $\mathbf{g} = -g\mathbf{k}$, we have

$$\nabla\left(\frac{\partial\phi}{\partial t} + H + gz\right) = 0$$

and therefore

$$\frac{\partial\phi}{\partial t} + \frac{p}{\rho} + \frac{1}{2}(\nabla\phi)^2 + gz = C(t) \tag{8.9}$$

The arbitrary function $C(t)$ can be absorbed into ϕ without affecting $\mathbf{u}$. Equation (8.9) is then an equation of pressure, the velocity field being obtained kinematically (without reference to the dynamical balance) by simply solving Laplace's equation subject to the *instantaneous* boundary conditions. The fact that the essential nonlinearity, as well as the time dependence of the equations, appears only in the pressure equation allows a rich family of solutions, many of them relevant to the problems of interest here.

8.2.1 *Examples*

The moving cylinder. Let a circular cylinder of radius a, axis perpendicular to the xy plane, move along the x axis with velocity $U(t)$. The boundary conditions, at the instant the center of the cylinder is at the origin, are then

$$\frac{\partial\phi}{\partial r} = U(t)\cos\theta \text{ when } r = a, \quad \phi \to 0 \quad \text{as } r \to \infty \tag{8.10}$$

An obvious harmonic function satisfying (8.10) is

$$\phi = -a^2U(t)r^{-1}\cos\theta \tag{8.11}$$

The velocity components and pressure are then given by

$$\begin{aligned} u_r &= a^2Ur^{-2}\cos\theta, \qquad u_\theta = a^2Ur^{-2}\sin\theta \\ \frac{p}{\rho} + gz &= a^2\dot{U}r^{-1}\cos\theta - \tfrac{1}{2}a^4r^{-4}U^2 \end{aligned} \tag{8.12}$$

The second term on the right of (8.12) is independent of θ and, at the surface of the cylinder, exerts no force. The first gives an x component of force equal to

$$F = \rho a^2 \pi \dot{U} = \text{hydrodynamic force exerted by cylinder on the fluid, per unit length} \quad (8.13)$$

Thus, apart from the force density needed to move the material of the cylinder with speed $U(t)$, and the force of buoyancy, there is an added hydrodynamic force due to the inertia of the fluid in which the cylinder moves ($-F$ is the inertial force). If $U = \dot{X}$, then (8.13) has the form $F = \rho a^2 \pi \ddot{X}$, so that $\rho a^2 \pi$ represents a certain mass density (per unit length of cylinder) that appears in Newton's law for the motion of the cylinder in a fluid environment. We define $\rho a^2 \pi$ to be the *apparent* (or added, or virtual) *mass* of the cylinder. This phenomenon, a direct result of the inertial force, is of course a familiar one to human swimmers. Rapid arm movement in water is hindered as if the mass had increased. We shall return to this subject later, because it is basic to propulsive mechanisms of fish.

Corner flows. The steady flow with potential $= U(x^2 - y^2)$ represents flow in a corner (Figure 8.2*a*). Note that a rectangular "parcel" of fluid remains rectangular and is compressed as shown. The absence of any "spin" of the parcel reflects irrotationality.

Vortex flow. Here $\phi = (\Gamma/2\pi) \arctan (y/x)$. The flow is irrotational except at the origin. A fluid element not at the origin rotates on a circle but maintains its orientation (see Figure 8.2*b*).

Figure 8.2. Streamlines of irrotational steady flows. (*a*) Corner flow. (*b*) Vortex flow.

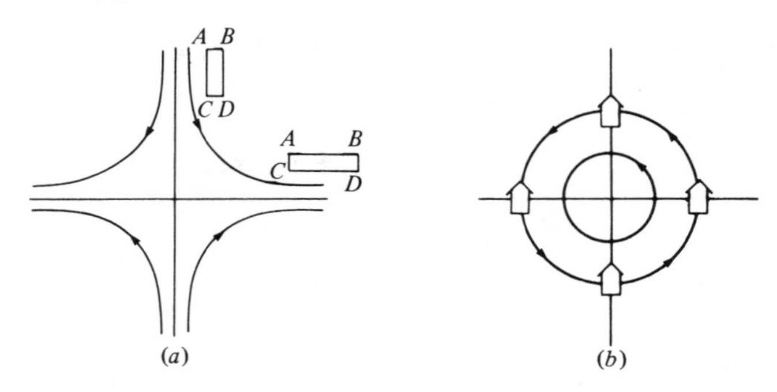

8.2.2 Complex variables

Two-dimensional irrotational flow can be studied by the methods of complex analysis. Let $z = x + iy$ be the complex variable and $w(z) = \phi + i\psi$ be any analytic function. Then from the Cauchy–Riemann equations, we see that ϕ and ψ define the velocity potential and stream function, respectively, for an irrotational flow. In the examples above, $w = -Ua^2z^{-1}$ (cylinder), $w = Uz^2$ (corner flow), and $w = -(i\Gamma/2\pi) \ln z$ (vortex flow).

More complicated flows can be built up by superposition or, frequently, through the application of a conformal mapping. Suppose that $w(\zeta)$ corresponds to a potential flow in the ζ plane ($\zeta = \xi + i\eta$) involving a boundary C where $\partial\phi/\partial n = 0$. Further, suppose that C is a simple closed curve (no self-intersections) and that the flow is exterior to C. Now, let $z = f(\zeta)$ map C onto another simple closed curve C' in the z plane such that exterior $(C) \rightarrow$ exterior (C'). As $w(\zeta) = \phi(\xi, \eta) + i\psi(\xi, \eta)$ represents a flow, it follows that C' is a streamline $\psi =$ constant (C' is the preimage of C under f^{-1}). Eliminating ζ, we have $W(z) = w(f^{-1}(z))$ as the complex potential for a flow in the z plane. Think of C' as a very complicated curve and C as a rather simple one to appreciate the power of the method.

Example. We consider flow past an ellipse (Figure 8.3). Let

$$z = f(\zeta) = \zeta + \frac{c^2}{4\zeta} \tag{8.14}$$

where c is real. This is a case of the *Joukowski transformation,* mapping the circle $\zeta = ae^{i\theta}$, $a > 0$, $0 \leq \theta < 2\pi$, onto the ellipse

$$x = \left(a + \frac{c^2}{4a}\right) \cos\theta \equiv \alpha \cos\theta$$

$$y = \left(a - \frac{c^2}{4a}\right) \sin\theta \equiv \beta \sin\theta$$

Figure 8.3. Conformal mapping to flow over an ellipse.

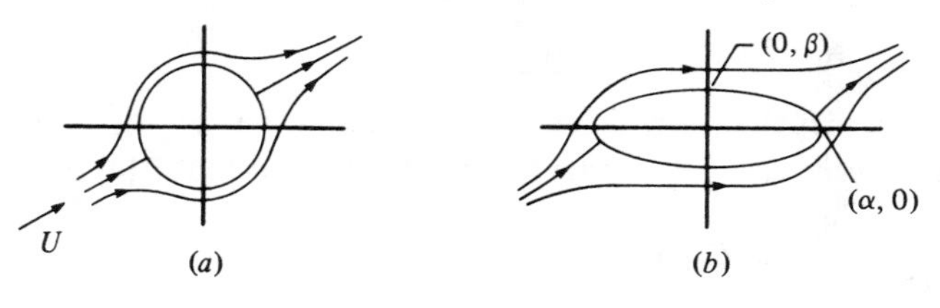

having major and minor semiaxes α and β. So we take (see the moving cylinder in Section 8.2.1)

$$w = U\left(\zeta e^{-i\lambda} + \frac{a^2}{\zeta} e^{+i\lambda}\right) \tag{8.15}$$

representing uniform flow at an angle of attack λ over a cylinder. Inverting (8.14) and making $z \sim \zeta$ for $|z|$ large, we obtain

$$\zeta = \tfrac{1}{2}(z + \sqrt{z^2 - c^2}) \tag{8.16}$$

The singular points at $z = \pm c$ are the foci of the ellipse and therefore in its interior. Note that if $c = 2a$ and $\lambda = \pi/2$, the flow above corresponds to the broadside-on flow of a fluid past a flat plate of width $4a$.

8.3 Translation of a solid

By translation of a solid body, we mean that each point of the body moves with the same velocity. The boundary condition (8.8) becomes [see (8.10)]

$$\frac{\partial \phi}{\partial n} = \mathbf{U} \cdot \mathbf{n} \qquad \text{on } S \tag{8.17}$$

in the stationary frame, where $\mathbf{U}(t)$ is the velocity of translation.

We consider the kinetic energy of the fluid, in two ($N = 2$) or three ($N = 3$) dimensions:

$$E(t) = \tfrac{1}{2}\rho \int (\nabla\phi)^2 \, dV, \qquad V = \text{exterior of } B \tag{8.18}$$

We shall suppose that this improper integral is convergent (absolutely); it is sufficient that ϕ decay like r^{-1} in three dimensions or is $r^{-\epsilon}$ ($\epsilon > 0$) in two dimensions. Now, since both Laplace's equation and (8.17) are linear in ϕ, we see that

$$\phi = \mathbf{U} \cdot \mathbf{\Phi} \tag{8.19}$$

where

$$\mathbf{n} \cdot \nabla\mathbf{\Phi} = \mathbf{n} \qquad \text{on } S$$

Since $\mathbf{\Phi}$ depends only on the instantaneous shape of the body, we see that E becomes independent of t if $\mathbf{U}$ is a constant. But generally

$$\frac{dE}{dt} = \mathbf{F} \cdot \mathbf{U}$$

where **F** is the force exerted by the fluid on the sphere. We have thus uncovered:

D'Alembert's paradox. If kinetic energy defined by (8.18) converges, the drag (the component of **F** parallel to **U**) vanishes in uniform translation. Of course, the result is not really paradoxical, because the sources of drag in the Eulerian realm need not be realized in the idealized model of irrotational flow.

We should note that, nevertheless, a *torque* is generally required to maintain uniform translation in an inviscid fluid. Also, in two dimensions the complex potential will often behave like $w = a \ln z + bz^{-1} + \cdots$ for large z. If Re $(a) = 0$, the energy can still be shown to exist [using the positivity of the integrand, apply the divergence theorem and evaluate $\phi(\partial\phi/\partial n)$ on a large circle whose radius tends to infinity]. But in this case there is a component of $-\mathbf{F}$ perpendicular to **U** (the *lift*), proportional to Im (a). Finally, if gravity is present, the result applies to $\mathbf{F} - \mathbf{F}_b$ when $\mathbf{F}_b$ is the buoyancy force, the work of the latter being realized as potential energy of the body.

We thus see that for a single translating solid, a force parallel to **U** arises only when **U** changes with time. One way to determine it would be to compute the total momentum

$$\rho \int \nabla\phi \, dV$$

but unfortunately, in the cases of interest, $\phi = O(r^{1-N})$ at infinity and the last integral is only conditionally convergent (see Exercise 8.5). To avoid this difficulty, we again work with the kinetic energy, now considering a region V bounded by S and a large surface Σ. We write

$$E = -\tfrac{1}{2}\rho \int (\mathbf{U} + \nabla\phi) \cdot (\mathbf{U} - \nabla\phi) \, dV + \tfrac{1}{2}\rho U^2 J(V) \tag{8.20}$$

where J is the content (area or volume) of V. By (8.17), $\mathbf{U} - \nabla\phi$ has zero normal component on S (this fact provides the main motivation for this method) and by writing the integrand in (8.20) as $\nabla \cdot (\mathbf{U} \cdot \mathbf{r} + \phi)(\mathbf{U} - \nabla\phi)$, the divergence theorem gives

$$E = \tfrac{1}{2}\rho \int \left(\frac{\partial\phi}{\partial n} - \mathbf{U} \cdot \mathbf{n}\right)(\mathbf{U} \cdot \mathbf{r} + \phi) \, d\Sigma + \tfrac{1}{2}\rho U^2 J(V) \tag{8.21}$$

We shall assume that

$$\phi = -\frac{\mathbf{A} \cdot \mathbf{r}}{r^N} + O(r^{-N}), \qquad r \to \infty \tag{8.22}$$

where **A** is some time-dependent vector. Taking Σ to be spherical, we may evaluate (8.21) using (8.22), giving

$$E = \pi(N - 1)\rho\mathbf{A} \cdot \mathbf{U} - \tfrac{1}{2}\rho J(B) U^2 \tag{8.23}$$

where B denotes the region of the body. Using (8.19), it follows that $\mathbf{A} = \mathbf{U} \cdot \mathbf{m}$ for some tensor **m** depending only on the shape of S, and therefore

$$E = \tfrac{1}{2}M_{ij}U_iU_j, \qquad M_{ij} = 2\pi\rho(N - 1)m_{ij} - \rho J(B)\delta_{ij} \tag{8.24}$$

We refer to **M** as the *apparent mass tensor* for the body.

The foregoing analysis of the apparent mass of solids may be found in Landau and Lifschitz (1959); it is interesting as an example of the computation of a physical property at a "surface at infinity." To take an example, for an ellipse one easily finds [using (8.14) and following]

$$\mathbf{M} = \pi\rho \begin{pmatrix} \beta^2 & 0 \\ 0 & \alpha^2 \end{pmatrix}$$

reducing to the result (8.13) given earlier when $\alpha = \beta$.

A *local* determination of **M** can also be useful (see Section 8.4):

$$M_{ij} = -\rho \int \Phi_i n_j \, dS \tag{8.25}$$

It can be shown (see Exercise 8.4) that **M** (and therefore **m**) is symmetric, as one would expect, although this is not obvious from (8.25).

Differentiating E and using (8.24) and the symmetry of **M**, we have

$$\frac{dE}{dt} = M_{ij}\dot{U}_iU_j = \mathbf{F} \cdot \mathbf{U} = \frac{d\mathbf{P}}{dt} \cdot \mathbf{U} \tag{8.26}$$

where by Newton's law **P** is the total linear momentum of the fluid. From (8.26) we see that

$$\mathbf{P} = \mathbf{M} \cdot \mathbf{U}$$

provided that body shape is independent of time. Here $-\mathbf{F}$ is the inertial force, **F** being the force that must be applied to the body to accelerate it.

8.4 Locomotion by recoil and squirming

We ask now how these results are changed if the body is deformable. Suppose that S is impermeable, that the content of B is constant, but that some internal mechanism deforms S as a periodic function of time, independently of whatever surface stresses may act. Also, we allow the internal mass distribution to vary independently of the surface shape.

When such an object is placed in an inviscid fluid at rest, will it swim? Some affirmative cases were given by Saffman (1967), one of which is easily visualized. Consider two dimensions and an elliptical body of variable eccentricity e but constant area. Let the body consist of light shell and supporting tissue, and suppose that it contains a concentrated mass; the latter can be moved along one of the principal axes. Then, as we indicate in Figure 8.4, the variation of apparent mass with e gives control over the distance of recoil accompanying any shift of the center of mass. By changing e out of phase with shifts of the mass, the body can displace itself with constant mean speed. Similar mechanisms are available for a homogeneous body, for there the deformation of S alters the momentum of body and fluid. Although the inertial forces are not now independent of the shape of S, Saffman exhibits deformations of a sphere having the desired properties.

To analyze these problems, it is convenient to consider the special case where centroid and center of mass lie on a given line for all time; this can be realized with deformable bodies having sufficient symmetry. We also again exclude gravity, because bouyancy will generate a moment about the centroid if the latter does not coincide with the center of mass. Let the instantaneous velocity of centroid and center of mass be $U(t)$ and $U(t) + \Delta U(t)$, respectively.

Figure 8.4. (*a*) and (*b*) Recoil swimming. (*a*) Mass shifts to right, body moves left a distance L_1. (*b*) Mass shifts to left, body moves right a distance $L_2 < L_1$. Over one period, the body moves $L_1 - L_2$ to the left. (*c*) Squirming of a deformable body.

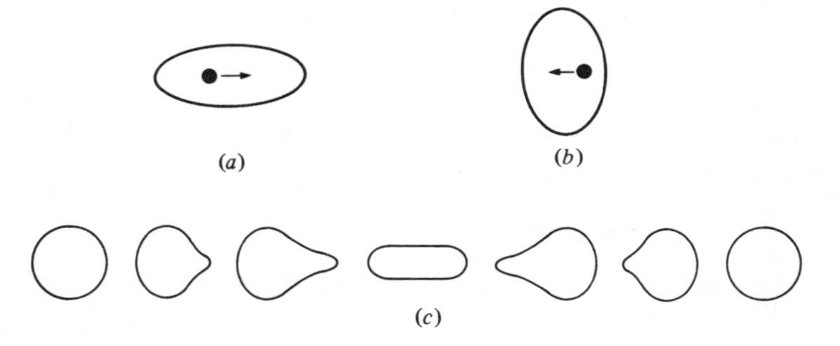

If the motion starts from rest, conservation of momentum implies that the total momentum (body and fluid) remains zero and we have

$$m(U + \Delta U) + P = 0 \tag{8.27}$$

where P is the fluid momentum and m the constant mass of B. We next decompose P:

$$P = M(t)U(t) + P_D(t), \qquad P_D = \mathbf{P}_D \cdot \mathbf{i} \tag{8.28}$$

where M is the apparent mass at time t for motion along the given line $\mathbf{U} = U\mathbf{i}$ and P_D is the excess momentum due to surface deformation. To find $P_D(t)$, it is convenient to compute the force in the stationary frame. To avoid convergence difficulties, we introduce a large fixed surface Σ containing S, and a potential ϕ^* equal to ϕ on S and 0 on Σ. Clearly, $\phi^* \to \phi$ as $\Sigma \to \infty$, and we may assume that $\Phi^* = O(r^{1-N})$. We now have ($\mathbf{n}$ outward on S and Σ)

$$\begin{aligned}
\mathbf{F}^* &= \int p^*\mathbf{n}\, dS \qquad \text{(see Chapter 1)} \\
&= -\rho \int \left[\frac{\partial \phi^*}{\partial t} + \tfrac{1}{2}(\nabla\phi^*)^2\right]\mathbf{n}\, dS \qquad \text{(Bernoulli's theorem)} \\
&= \rho \int \left[\frac{\partial}{\partial t}\nabla\phi^* + \nabla \cdot (\nabla\phi^*\nabla\phi^*)\right] dV - \frac{\rho}{2}\int (\nabla\phi^*)^2\mathbf{n}\, d\Sigma \\
&= \rho \int \frac{\partial \nabla\phi^*}{\partial t}\, dV - \rho \int \nabla\phi^*(\nabla\phi^* \cdot \mathbf{n})\, dS \\
&\quad + \frac{\rho}{2}\int [\nabla\phi^*(\nabla\phi^* \cdot \mathbf{n}) - \tfrac{1}{2}(\nabla\phi^*)^2\mathbf{n}]\, d\Sigma \\
&= \frac{\partial}{\partial t}\rho \int \nabla\phi^*\, dV + \Sigma \text{ boundary terms} \qquad \text{[by (1.11)]} \\
&= -\frac{\partial}{\partial t}\rho \int \phi^*\mathbf{n}\, dS + \cdots
\end{aligned}$$

As the terms over Σ converge to zero as Σ tends to infinity, we get

$$\mathbf{P} = -\rho \int \phi\mathbf{n}\, dS$$

With $\phi = \mathbf{U} \cdot \mathbf{\Phi} + \phi_D$, we have thus established (8.25) for a general time-dependent boundary, and obtain

$$P_D = -\rho \int \phi_D(\mathbf{n} \cdot \mathbf{i})\, dS$$

We see that dP_D/dt is the force that must be applied to the body to keep its centroid stationary while the surface deforms. The force is periodic in t if the surface deformation is periodic; hence, P_D also is periodic. Consequently, *in irrotational inviscid flow, the inertial forces from periodic body deformation contribute no net momentum to the fluid.* In other words, the mean acceleration will vanish. There emerges here a crucial distinction between inertial and vortex forces (Chapter 9), for in rotational inviscid flow, body deformation can generate a mean *thrust.*

If there is no body deformation, (8.27) and (8.28) combine to give

$$\langle U \rangle = -\left\langle \frac{m\,\Delta U}{m + M} \right\rangle$$

and the average can be made nonzero by constructions such as the one shown in Figure 8.4. If the body is homogeneous, we have similarly

$$\langle U \rangle = -\left\langle \frac{P_D}{m + M} \right\rangle$$

and again, as Saffman (1967) has shown, examples can be found giving a nonzero average. (See Figure 8.4c.)

Recoil and squirming locomotion are thus examples of mechanisms depending solely on inertial forces. Whereas all Eulerian swimmers utilize the inertia of the fluid, natural swimming tends to be dominated by the effects associated with vorticity. The resulting inviscid calculation of thrust would then determine a swimming speed provided viscous drag can be computed separately. If an unambiguous decomposition into thrust and drag were possible in the present examples, zero thrust would then imply that in a real fluid of small viscosity an Eulerian squirmer could not swim! Note, however, that at *low* Reynolds number the squirming movement of Figure 8.4*c* may succeed since it is not time reversible (see also Exercise 8.6). It is, therefore, possible that squirming is also realized at high Reynolds number, but the mechanism may depend in an essential way upon the viscosity.

Exercises

8.1 A flow with zero vortex force is called a Beltrami field. Show that $\mathbf{u} = \nabla \times \nabla \times \mathbf{A} + \alpha\nabla \times \mathbf{A}$ is such a field for any $\mathbf{A}$ satisfying $\nabla^2\mathbf{A} + \alpha^2\mathbf{A} = 0$.

8.2 Derive the first of (8.10) from the equation for S and (8.8).

8.3 Find the potential flow due to rectilinear motion of a sphere of radius a. Evaluate the pressure force and check your expression for apparent mass against the general result in (8.24).

8.4 Show that $M_{ij} = M_{ji}$ [see (8.25)] by considering $\int (\Phi_j n_i - \Phi_i n_j)$ dS, using the boundary condition on S to write the integral in the form

$$\int \left(\Phi_j \frac{\partial \Phi_i}{\partial x_k} - \Phi_i \frac{\partial \Phi_j}{\partial x_k} \right) n_k \, dS$$

Then apply Green's theorem to the external domain with $\boldsymbol{\Phi} = O(r^{1-N})$ at infinity.

8.5 The fluid momentum might be defined by $\rho \int \nabla \phi \, dV$. With $\phi = -(\mathbf{A} \cdot \mathbf{r}) r^{-N} + O(r^{-N})$, the integral is, however, only conditionally convergent and depends on the shape of Σ. Show that an erroneous expression for apparent mass is obtained by taking Σ to be spherical, writing the integral in the form

$$\rho \int \nabla \cdot (\mathbf{r} \cdot \nabla \phi) \, dV$$

and using

$$(\mathbf{r} \cdot \mathbf{U}) \, (\mathbf{U} \cdot \mathbf{n}) \, dS = U^2 J(B)$$

8.6 (Suggested by Charles Peskin) Show that a homogeneous squirmer with time-reversible boundary motion (see Chapter 2) will not swim.

9

The Eulerian realm: the vortex force

9.1 The kinematics of vorticity

We study now the primary mechanisms by which lift and thrust are generated at high Reynolds numbers. To treat this topic, we need some fundamental results concerning the *vorticity* field of a rotational perfect fluid flow. Given a fluid flow with velocity field $\mathbf{u}$, we recall that the vorticity field is defined by $\omega = \nabla \times \mathbf{u}$. To obtain an equation involving only $\mathbf{u}$ and ω, we take the curl of (8.2) and use (8.3) and the vector identity

$$\nabla \times (\mathbf{A} \times \mathbf{B}) = \mathbf{B} \cdot \nabla\mathbf{A} + \mathbf{A}\nabla \cdot \mathbf{B} - \mathbf{A} \cdot \nabla\mathbf{B} - \mathbf{B}\nabla \cdot \mathbf{A}$$

This yields the *vorticity equation*,

$$\frac{\partial \omega}{\partial t} + \mathbf{u} \cdot \nabla\omega = \omega \cdot \nabla \mathbf{u} \tag{9.1}$$

Using the notation introduced in Section 1.2, we may also write (9.1) in the form

$$\frac{d\omega}{dt} = \omega \cdot \nabla\mathbf{u}$$

where we now refer to $d\omega/dt = \partial\omega/\partial t + \mathbf{u} \cdot \nabla\omega$ as the *material derivative* of ω. We note from the vorticity equation that an observer moving with the fluid will see the vorticity change in response to gradients of $\mathbf{u}$.

Our first aim will be to understand how the vorticity field evolves in a fluid flow. For this it will be helpful to momentarily regard $\mathbf{u}$ in (9.1) as a *known* vector field, in which case (9.1) may be thought of as a linear equation for ω. Of course, ω and $\mathbf{u}$ are closely related because the former is the curl of the latter, and it must turn out that (9.1) preserves this connection. That is, if $\mathbf{u}$ is given and we solve (9.1) for a vector field ω, equal to $\nabla \times \mathbf{u}$ when $t = 0$, we should find that $\omega = \nabla \times \mathbf{u}$ for all t, at least under suitable conditions on $\mathbf{u}$. For our purposes what is important is to recognize that much can be learned about the vorticity field without considering the dynamical origin of the underlying velocity field. These considerations comprise the *kinematical* theory of vorticity.

The kinematics of vorticity can be given a geometrical interpretation in terms of *material invariants* of the flow **u**. The simplest material invariant is a scalar quantity that is fixed to a fluid particle. An example of such a scalar invariant would be temperature, provided that the thermal diffusivity of the material is zero. Since a scalar invariant ϕ does not change as we follow a fluid particle, it will satisfy the equation

$$\frac{d\phi}{dt} = 0$$

As is clear from (9.1), the vorticity field does *not* consist of invariant vectors, satisfying $d\omega/dt = 0$. The proper geometrical interpretation lies in the closely related concept of a *material line.* Imagine a curve in space drawn at $t = 0$, then follow the evolution of the line under the flow **u.** In general, the length as well as the shape of the line will be changed by the flow. We can describe this mathematically by considering the change in a differential line element **dl** during a short interval of time Δt (Figure 9.1). We see that

$$\mathbf{dl}(t + \Delta t) = \mathbf{dl}(t) + \mathbf{u}(\mathbf{r} + \mathbf{dl}(t))\,\Delta t - \mathbf{u}(\mathbf{r}, t)\,\Delta t + O((\Delta t)^2)$$

or

$$\frac{\mathbf{dl}(t + \Delta t) - \mathbf{dl}(t)}{\Delta t} = \mathbf{dl} \cdot \nabla\mathbf{u} + O(\Delta t)$$

Figure 9.1. Evolution of a material line.

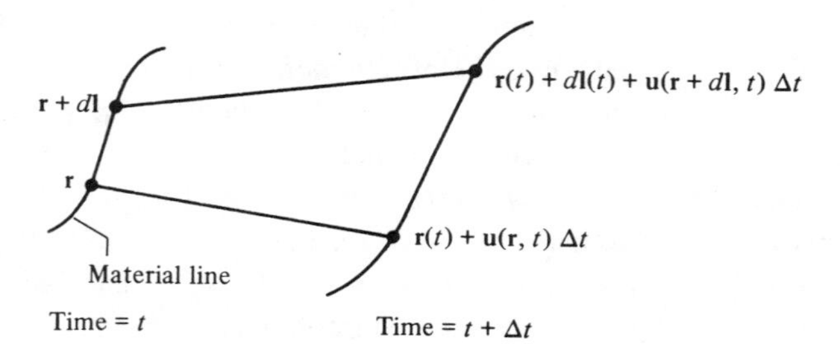

Hence,

$$\frac{d}{dt}(\mathbf{dl}) = \left(\frac{\partial}{\partial t} + \mathbf{u} \cdot \nabla\right) \mathbf{dl} = \mathbf{dl} \cdot \nabla \mathbf{u} \tag{9.2}$$

The term on the right of (9.2) describes the stretching and rotating of the line element.

If we imagine space to be filled by oriented material lines, we can consider a small tube whose boundary consists of material lines. Then, as the fluid is incompressible, the *content* of the tube is invariant, and its cross-sectional area will change in response to stretching (Figure 9.2).

Comparing (9.1) and (9.2), we have the remarkable result that vorticity and the line element **dl** (think of the latter as attached to a given material point) satisfy the same evolution equation. So, if we assume that the evolution equation uniquely determines ω (given its initial values), we have

$$\frac{\omega(\mathbf{r}, t)}{|\omega(\mathbf{r}_0, 0)|} = \frac{d\mathbf{l}(t)}{|d\mathbf{l}(0)|} \tag{9.3}$$

where $d\mathbf{l}(0)$ is the line element parallel to $\omega(\mathbf{r}_0, 0)$ at the point $\mathbf{r}_0 = \mathbf{r}(0)$.

Thus, the trajectories of the vorticity field have been shown to be material lines. An infinitesimal *vortex tube* can be considered as in Figure 9.2; it follows from (9.3) and the invariance of content of vortex tubes that $|\omega|A$ *is a scalar material invariant,* where A is the cross-sectional area of the tube. Thus, in particular, vorticity can be increased by the stretching of a vortex tube.

9.1.1 *Kelvin's circulation theorem*

This result will prove useful to us. Imagine an oriented simple closed material curve C. The *circulation* about C is defined by

$$\Gamma(t) = \oint_{C(t)} \mathbf{u}(\mathbf{t}) \cdot d\mathbf{l}(t) \tag{9.4}$$

Figure 9.2. Tube of material lines: $lA = l'A'$.

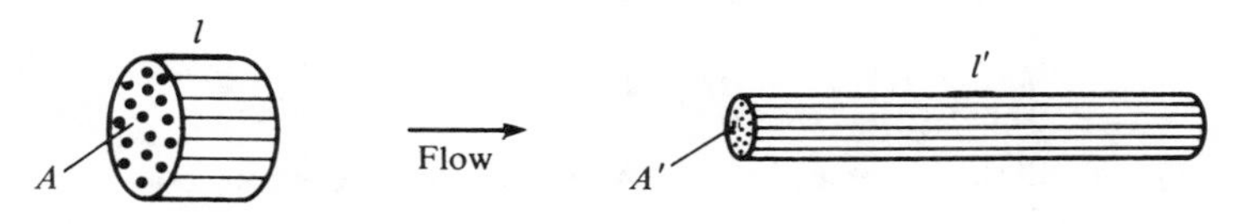

The material derivative of $\Gamma(t)$ is [using (9.2) and Euler's equation]

$$\begin{aligned}\frac{d\Gamma}{dt} &= \frac{d}{dt}\oint_C \mathbf{u}\cdot d\mathbf{l} = \oint_C \frac{d\mathbf{u}}{dt}\cdot d\mathbf{l} + \oint_C \mathbf{u}\cdot\frac{d\mathbf{l}}{dt} \\ &= \oint_C\left(-\frac{\nabla p}{\rho}\cdot d\mathbf{l} + \oint_C \mathbf{u}\cdot(d\mathbf{l}\cdot\nabla\mathbf{u}\right) \\ &= \oint_C \nabla\left(\tfrac{1}{2}u^2 - \frac{p}{\rho}\right)\cdot d\mathbf{l}\end{aligned}$$

As the last integrand is a perfect differential, we have

$$\frac{d\Gamma}{dt} = 0 \tag{9.5}$$

That is, Euler's equations conserve circulation about material curves, which is Kelvin's theorem. Note that if C is the boundary of an oriented surface S, Stokes' theorem implies that

$$\Gamma = \int_S (\nabla\times\mathbf{u})\cdot\mathbf{n}\,ds = \int_S \omega_n\,ds \tag{9.6}$$

and thus Γ measures the "flux" of vorticity through the surface S. Locally, this reduces to the invariance of $|\omega|A$ noted earlier. An immediate consequence of (9.6) is that (because by definition there is no flux through the sides of a tube) $|\omega|A$ is the same at all cross sections. This invariant of a vortex tube is known as its *strength*.

9.1.2 *Lagrangian description*

If one knows the Lagrangian coordinates $\mathbf{r}(t;\mathbf{r}_0)$ of material points (recall that $\mathbf{r}$ is the position at time t of the particle initially at $\mathbf{r}_0$), then if $d\mathbf{l}$ is at $\mathbf{r}$ and $d\mathbf{l}(0)$ at $\mathbf{r}_0$, we have

$$d\mathbf{l}(t) = \frac{\partial\mathbf{r}}{\partial\mathbf{r}_0}\cdot d\mathbf{l}(0) \equiv \mathbf{D}\cdot d\mathbf{l}(0)$$

say. Using (9.3), we then obtain

$$\omega(\mathbf{r}(t,\mathbf{r}_0),t) = \mathbf{D}\cdot\omega(\mathbf{r}_0,0)$$

This explicitly "solves" (9.1) (provided that we know **D**).

9.1.3 *Circulation in irrotational flow*

It is a disturbing fact that boundary-value problems for Euler's equations are not generally well posed, usually because of the incompleteness of boundary conditions. For example, the steady two-dimensional irrotational flow of a fluid over a fixed rigid circular cylinder of radius a, satisfying the condition that normal velocity vanish on the surface, was seen in Chapter 8 to be given by

$$\mathbf{u} = \mathbf{U} + a^2\nabla \frac{\mathbf{U} \cdot \mathbf{r}}{r^2} = \mathbf{u}_1 \tag{9.7}$$

However, another solution satisfying the same boundary conditions is

$$\mathbf{u} = \mathbf{u}_1 + \frac{\Gamma}{2\pi} \nabla\theta \tag{9.8}$$

where θ is the polar angle (measured positive counterclockwise, say) and Γ is a constant. The solutions differ by the addition of the vortex flow (center at the origin) considered in Chapter 8. Furthermore, whereas (9.7) gives zero circulation on the surface, (9.8) gives

$$\oint_{r=a} \mathbf{u} \cdot d\mathbf{r} = \Gamma \tag{9.9}$$

Now, using (9.5), we see that circulation is zero on any curve spanned by a surface S having a neighborhood where $\mathbf{u} = \nabla\phi$. Thus, the circulation on any curve winding once about the cylinder is Γ. (In terms of the complex potential, these results follow from Cauchy's theorem and the residue at $z = 0$.)

Also, whereas (9.7) gives zero force on the cylinder, evaluating the pressure on $r = a$ shows that with (9.8) the force is

$$\mathbf{F} = \frac{a\rho}{2} \int_0^{2\pi} u^2 (\mathbf{i} \cos\theta + \mathbf{j} \sin\theta)\, d\theta = -\rho U\Gamma\mathbf{j} \tag{9.10}$$

The meaning of this lift force ($\mathbf{U} = U\mathbf{i}$) can be clarified by computing it over a distant contour C using conservation of momentum and Bernoulli's theorem. With $\mathbf{u} - \mathbf{U} = \mathbf{v}$, we get [see the text following (8.28)]

$$\begin{aligned} \mathbf{F} &= \rho \int_C [\mathbf{U}(\mathbf{v} \cdot \mathbf{n}) + \mathbf{v}(\mathbf{U} \cdot \mathbf{n}) - (\mathbf{U} \cdot \mathbf{v})\mathbf{n}]\, ds \\ &= \rho\mathbf{U} \times \int_C (\mathbf{v} \times \mathbf{n})\, ds = \rho\mathbf{U} \times \mathbf{k} \int_C \mathbf{v} \cdot d\mathbf{l} \end{aligned} \tag{9.11}$$

where **k** is out of the plane. Thus, it is the circulation, or the net flux of vortex lines within the cylinder across the plane of the flow, that generates the lift. The result (9.10), for steady irrotational two-dimensional flow with circulation Γ, is the *Kutta–Joukowski theorem.*

Vorticity "within" a body is called *bound vorticity.* In practice, bound vorticity is always deduced by calculating circulation in the domain of the fluid, but once the structure of the flow field is known in terms of singular solutions of Laplace's equations, the body surface may be ignored and the singularity, associated with (infinite) vorticity having total strength $2\pi\Gamma\mathbf{k}$, generates a lift (per unit length) $\rho\mathbf{U} \times 2\pi\Gamma\mathbf{k}$. Note that two dimensions is special in that lift is possible even though vortex lines lie fully within the body. In three dimensions there is no reason why a vortex line cannot penetrate a boundary but, owing to the solenoidal property of a vortex line that does not close, it cannot simply terminate (except at infinity). On the other hand, a *closed* bound vortex tube of strength Γ will generate the force

$$\rho\Gamma\mathbf{U} \times \int_{\text{tube}} \mathbf{dl} = 0$$

Thus, in three-dimensional steady flow, lift can be generated only by vortex tubes that terminate at infinity, the bound vorticity then being equal to the portion within the body (see Figure 9.3 and Exercise

Figure 9.3. Bound vorticity in steady flow. Vortex tubes oriented by right-hand rule. (*a*) Two dimensions, irrotational flow. (*b*) Three dimensions.

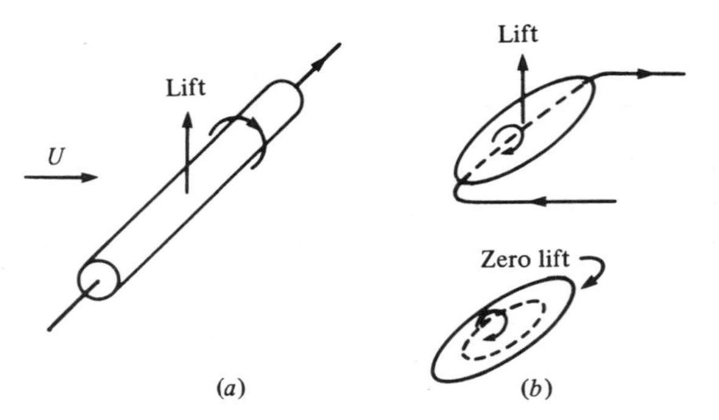

9.4).† But the flow exterior to S is no longer irrotational. The free vorticity is usually referred to as the *shed vorticity.*

9.2 The role of viscosity

The nonuniqueness exhibited above reveals a basic defect in perfect flow theory based solely upon the boundary condition $dS/dt = 0$. In two dimensions, the nonuniqueness can be "removed" by taking equivalence of circulation about bodies as an added constraint. But there are other, analogous problems in three dimensions that are not so easily resolved. As it seems unlikely that the full Navier–Stokes equations with the condition of adherence would have these defects, it must be recognized that viscosity, however small, must play an essential role in "selecting" appropriate Euler flows.

We turn, therefore, to observations of real flows. Two-dimensional flow over a circular cylinder can be approximated in a wind tunnel, and it is found that at high Reynolds numbers the flow is not steady and, owing to a complicated wake region, is not at all close to the ideal limit (9.7) [see, e.g., the photographs in Batchelor (1967)]. For more slender bodies, the ideal limits (9.7) and (9.8) are, however, useful. Suppose, for example, that a body such as an airfoil is accelerated from rest in a fluid initially at rest. According to Kelvin's theorem, the circulation about a material curve around the body will remain zero. If initial acceleration is sufficiently rapid, it is found that the ideal flow with zero circulation is set up initially, but if the trailing edge is sufficiently sharp, the very low pressure associated with the flow there cannot be sustained as the boundary layer begins to develop. What happens ultimately is shown in Figure 9.4. A tube of concentrated vorticity is shed from the trailing edge, and a smooth flow across the trailing edge is thereby established. Although circulation on the original material curve remains zero, there is now circulation about the airfoil. The condition of zero singularity at the trailing edge is known as the *Kutta condition.* It is in part the sharp trailing edge of an airfoil that forces this sequence of events and the generation of lift. We study the process in more detail in Chapter 11.

†This conclusion assumes that in the steady flow all fluid particles initially upstream of the body eventually lie arbitrarily far downstream.

In free swimming and flying, the situation regarding shed vorticity is complicated by the unsteadiness of the flow and will be discussed below in the context of specific applications. However, the Euler flow approximation tends to be more closely realized because of the absence of mean force as well as the unsteadiness of the flow. For then, over one cycle of the motion, cumulative effects of viscosity associated with boundary-layer growth and flow separation may not have sufficient time to develop.

9.2.1 *An example*

We now give a concrete example of the use of the Kutta condition. Consider a lifting flat piece of chord $4a$. From the results of Chapter 8, we have

$$z = f(\zeta) = \zeta + \frac{a^2}{\zeta}$$

$$w(\zeta) = U\left(e^{-i\lambda}\zeta + \frac{a^2 e^{i\lambda}}{\zeta}\right) - \frac{i\Gamma}{2\pi}\ln\zeta$$

Figure 9.4. Development of lift in acceleration from rest. (a) $t = 0+$, no lift. (b) $t > 0$, airfoil experiences lift.

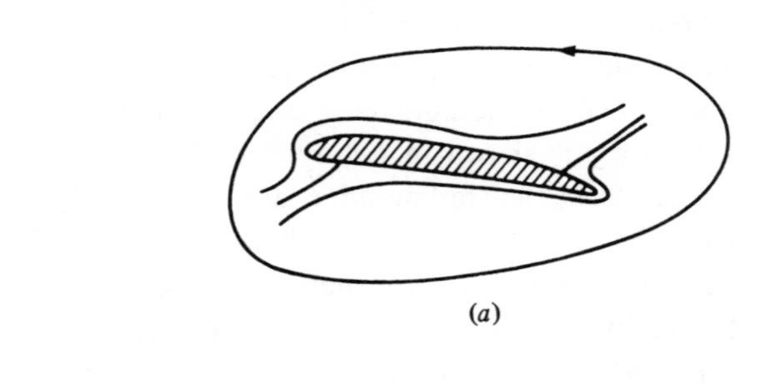

(a)

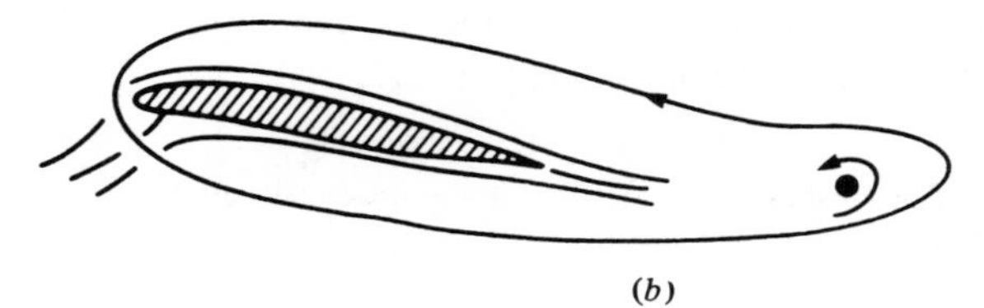

(b)

where the vortex flow is added to meet the Kutta condition. Thus, to determine Γ, we set

$$w'(\zeta) = 0 \qquad \text{at } \zeta = a$$

in order to place the stagnation point at the preimage of the trailing edge. This gives

$$\Gamma = 4\pi a U \sin \lambda$$

Thus from (9.10) the lift is

$$L = 4\pi\rho U^2 a \sin \lambda \tag{9.12}$$

Note that the lift is not perpendicular to the plate but to the free stream. The apparent contradiction is resolved by analyzing the vicinity of the leading edge. It is found that the low-pressure singularity contributes a finite thrust, called the *leading edge suction force.* Of course, (9.12) cannot be used if $\lambda = \pi/2$; in fact, viscous effects cause the airfoil to stall when $\lambda \sim 15°$.

Exercises

9.1 Show that $\omega = \mathbf{k} \cdot \omega$ is a material invariant in two-dimensional flow in the plane perpendicular to **k.**

9.2 Let $d\mathbf{s}$ be a small-area element of a material surface. Show by writing $d\mathbf{s} = d\mathbf{l}_1 \times d\mathbf{l}_2$ that

$$\frac{d(d\mathbf{s})}{dt} = d\mathbf{s} \cdot \nabla \mathbf{u}$$

9.3 Verify (9.10).

9.4 Show that the lift force produced by a vortex tube of strength $2\pi\Gamma$ bound to a three-dimensional solid body and crossing S at points P and Q (Figure 9.3*b*) is

$$\rho\Gamma\mathbf{U} \times \mathbf{l}$$

where $\mathbf{l}$ is the projection of the line PQ onto the plane normal to $\mathbf{U}$.

10

The swimming of fish

We now take up the mechanical principles underlying Eulerian swimming of a thin, flexible creature, with the aim of understanding morphology in terms of the mechanisms of propulsion. For summaries of the related biology see Lighthill (1975, chap. 2) and Wu (1971 *a*).

10.1 Small-perturbation theory of slender fish

Many fish change shape rather gradually along the anterior–posterior axis. (There are, however, many exceptions: for example, angelfish.) As a first approximation, we consider the swimming of slender, neutrally bouyant fish. By slender we mean, among other things, that the cross-sectional area of the body changes slowly along its length. Notation and terminology are summarized in Figure 10.1.

Necessary conditions for the validity of slender-body theory are such as to ensure that velocity perturbations caused by the fish are a small fraction of its swimming speed. It is certainly sufficient that the body surface S be smooth and that tangent planes always make a small angle with the x axis. However, it turns out that if these conditions are exactly met and the cross-sectional area is zero at the extremities, the theory predicts zero mean thrust. Fortunately, the model also allows us to treat "slender" fish with sharp downstream edges, at the caudal fin, for example, even though $s(x)$ may be discontinuous there. Also, it is not necessary that surface slope always be small, at the nose of the fish, for example, provided that this occurs over at most a fraction of its length.†

We restrict our study to a body with the following properties: (1) When "stretched straight" it is laterally symmetric, this being a property of many fish; (2) with the exception of the vicinity of the nose and a vertical edge of the caudal fin, surface slopes are small and the body

†Note that slender-body theory fails at the large dorsal fin of many fish because the slope of the *upstream* edge is not small.

surface is smooth (the addition of midbody fins is considered in Section 10.4); and (3) the cross-sectional area is zero at both ends, the shape of each section being arbitrary otherwise.

Movements are assumed to be lateral, and in the small-amplitude theory we require that

$$\left|\frac{\partial h}{\partial x}\right| \ll 1, \qquad \left|\frac{\partial h}{\partial t}\right| \ll U \tag{10.1}$$

where U is the swimming speed (in the direction of negative x). These assumptions ensure that flow perturbations are small, so that in the comoving frame

$$\frac{d}{dt} \simeq \frac{\partial}{\partial t} + U\frac{\partial}{\partial x} = D \tag{10.2}$$

From the point of view of the dynamical balance, (10.2) carries with it the implication that a *stationary* observer will see a thin (compared to body length) slice of water normal to the body axis pass down the length of the fish with speed U. The time dependence of the body cross section intercepted by the slice will generate locally an approximately two-dimensional, time-dependent flow field of the kind studied in Chapter 8. There is one new feature – that the *area* of the cross sec-

Figure 10.1. Notation for slender fish. (*a*) "Reality." (*b*) Slender fish without midbody fins. $|s'(x)| \ll 1$, $|\partial h/\partial x| \ll 1$, $|\partial h/\partial t| \ll U$.

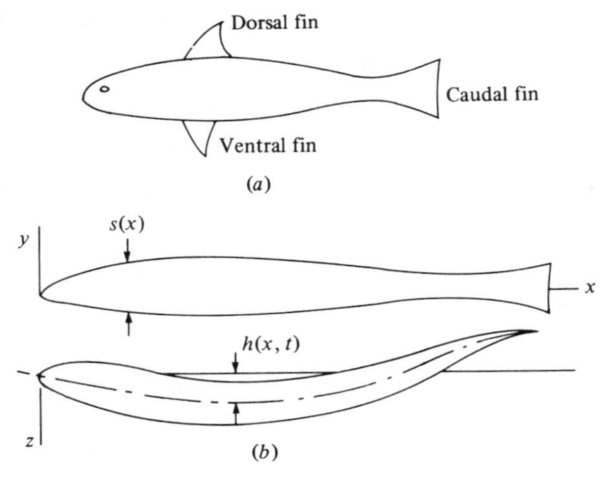

tion will also change. However, by lateral symmetry the area changes produce no z momentum and need not be considered in the lateral momentum balance.

We turn now to Lighthill's (1975, chap. 4) pioneering analysis of the foregoing problem. His small-amplitude theory is based upon a division of the rate of working of an Eulerian swimmer into a part $W_L(t)$ equal to the rate of working by the lateral motions of the body, and a part $-UT$ representing the effect of motion with speed U of a body delivering mean thrust T. The lateral working is readily found from direct computation of the inertial forces, but the thrust contribution is more accessible through an indirect energy balance. Recall that in a viscous medium the thrust and drag will balance, and to the extent that our inviscid theory holds in a real fluid, the work W_L will be the same as the exact swimming effort W_S.

The z component of the velocity of a cross section seen by the moving water slice (see above) is the approximate material derivative of the displacement $h(x, t)$:

$$w(x, t) = \frac{\partial h}{\partial t} + U\frac{\partial h}{\partial x} = Dh \tag{10.3}$$

If $m(x)$ is the apparent mass of the cross section at x, the lateral force exerted by the body on the water slice is the material derivative of mw.†

$$F_z = D(mw) \tag{10.4}$$

Using (10.3) and (10.4), the rate of working of the lateral motions is therefore, if l is the length of the fish,

$$\begin{aligned} W_L(t) &= \int_0^l F_z \frac{\partial h}{\partial t}\,dz = \int_0^l \frac{\partial h}{\partial t} D(mw)\,dx \\ &= \int_0^l D\left(mw\frac{\partial h}{\partial t}\right)dx - \int_0^l mw\frac{\partial}{\partial t}(Dh)\,dx \\ &= \frac{\partial}{\partial t}\int_0^l \left(mw\frac{\partial h}{\partial t} - \frac{1}{2}mw^2\right)dx + \left[Umw\frac{\partial h}{\partial t}\right]_{x=l} \end{aligned} \tag{10.5}$$

†For simplicity we take m to be independent of time, even though it is known that certain fish can vary the depth of the caudal fin as they swim.

where now $m(0) = 0$. We see from (10.5) that the *mean* rate of working depends only upon conditions at the posterior end:

$$\langle W_L \rangle = U\langle m(l)w(l, t)h_t(l, t)\rangle \qquad (10.6)$$

This situation is very different from anguilliform propulsion, where essentially all cross sections contribute. Among fish, the almost universal occurrence of a well-developed caudal fin can be taken as evidence in favor of (10.6).

The second stage of the argument utilizes the energy balance. The rate of working $W_L - UT$ must appear as kinetic energy of water slices intercepting the body:

$$W_L(t) = UT + \int_0^l D(\tfrac{1}{2}\, mw^2)\, dx \qquad (10.7)$$

where we use the relationship between kinetic energy and apparent mass given in Chapter 8. We can write (10.7) in the form

$$W_L(t) = UT + \frac{\partial}{\partial t}\int_0^l \tfrac{1}{2}mw^2\, dx + \left[\frac{1}{2}Umw^2\right]_{x=l} \qquad (10.8)$$

The second term on the right of (10.8) is the instantaneous rate of change of the kinetic energy generated by lateral movements; the last term accounts for the energy shed into the wake at the caudal fin (see below). Comparing (10.5) and (10.8) and again using (10.3), we have

$$T = m(l)[wW - \tfrac{1}{2}w^2]_{x=l} - \frac{\partial}{\partial t}\int_0^l mw\frac{\partial h}{\partial t}\, dx \qquad (10.9)$$

where $W = \partial h/\partial t$. Therefore,

$$\langle T \rangle = m(l)\langle [wW - \tfrac{1}{2}w^2]_{x=l}\rangle \qquad (10.10)$$

Thus, mean thrust, although it may be realized by pressure forces elsewhere on the body, is fully determined by conditions at the edge of the caudal fin.

We shall suppose that the purely *reactive* thrust (10.10) will balance *resistive* drag in steady swimming. Thus, from (10.10) we see that motion can occur only if w is positively correlated with $W - \tfrac{1}{2}w$.

Note that this implies that $\langle h_t^2 \rangle$ must exceed $U^2 \langle h_x^2 \rangle$. If motions were due to a progressive wave

$$h = -\frac{W}{V} \sin (x - Vt - l) \qquad (10.11)$$

then $w(l, t) = (V - U) WV^{-1} \cos (Vt)$ and

$$\langle T \rangle = \frac{m(l)}{4} W^2 \left(1 - \frac{U^2}{V^2}\right)$$
$$= \frac{m(l)}{4} H^2 (V^2 - U^2)$$

where $H = W/V$ is the maximum lateral excursion of the tail. Consequently, as in the Gray–Hancock theory of flagellar propulsion, the wave speed seen by the comoving observer must exceed the swimming speed for positive thrust. These requirements are supported by casual observations of swimming fish.

An appropriate efficiency of swimming is

$$\eta = \frac{U\langle T \rangle}{\langle W_L \rangle} = 1 - \frac{\langle W_L \rangle - U\langle T \rangle}{\langle W_L \rangle} \qquad (10.12)$$

Using (10.6) and (10.10), we get

$$\eta = 1 - \frac{1}{2} \frac{\langle w^2 \rangle}{\langle wW \rangle} \qquad (10.13)$$

Thus, to have high efficiency and positive thrust simultaneously, the fish should positively correlate w and W while keeping w as small as possible compared to W, given the thrust requirement (see Figure 10.2). The last constraint is an important one; for the wave motion (10.11), for example,

$$\eta = \frac{U + V}{2V}$$

so swimming is most efficient when thrust is zero.

If ϵl is taken as an amplitude of lateral movement and if a typical frequency is U/l, then the mean thrust predicted by (10.10) is $O(m(l) U^2 \epsilon^2)$, a rather small value. One could try to explain the quadratic dependence on ϵ by the small streamwise inclination of small inertial forces, but such an argument applies as well to bodies without caudal fins, which produce no thrust to this order. The explanation appears to be that the thrust is actually in part the mean of small streamwise components of small *vortex* forces. This can be visualized most readily for flat fish representable by a planar distribution of vorticity, with vortex shedding at the caudal fin edge. The vortical wake as well as the bound vorticity is involved, but for the moment consider only the bound vorticity. It is then not difficult to show [see, e.g., Wu (1971*b*) and Section 10.2] that the total strength of y vorticity in any section is just

$$\frac{\partial}{\partial x}(mw)$$

The speed w of the section then produces a vortex thrust

$$w\frac{\partial(mw)}{\partial x}$$

On the other hand, the inertial force is predominately but not exactly lateral, there being a contribution to thrust due to the body slope given by

$$-h_x D(mw)$$

Adding these contributions and integrating, we obtain

$$\int_0^l \left[w\frac{\partial(mw)}{\partial x} - h_x D(mw) \right] dx = [mwW]_{x=l} - \frac{\partial}{\partial t}\int_0^l mw\frac{\partial h}{\partial x}\,dx$$

giving us two terms of (10.9). The missing term describes the drag realized from the effect of wake vorticity, manifested by the continual creation of kinetic energy in the fluid downstream of the fish. (A close analog is the induced drag of finite lifting wings; see Chapter 11.)

Figure 10.2. Fin slope and speed reach a maximum simultaneously.

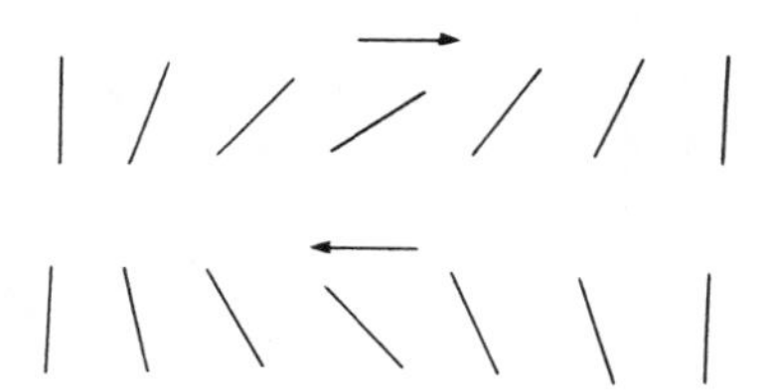

10.2 Vortex shedding

We study the flow downstream of the caudal fin with the aim of understanding the contribution of the vortical wake to the dynamical balance just described. It should first be noted that in the immediate vicinity of the fin edge, within one fin depth of the edge, say, the quasi two-dimensional analysis of inertial force breaks down, at least in principle. However, at the fin edge, we expect to find that a version of the Kutta condition of Chapter 9 should be applied to obtain physically reasonable flow. If this is done, it turns out that the troublesome region, which, after all, covers only a negligible fraction of the body length, can be "crossed" through the use of Kelvin's theorem, the latter providing in effect a matching condition between body and wake.

The principles underlying the Kutta condition require smooth flow off an edge, and bounded pressure. Inspection of the trailing edge of the lifting flat plate of Chapter 9 shows that if the angle of attack is small, the x component of the perturbation velocity vanishes there with error of order of the square of the angle of attack. In general, for time-dependent edges the condition at the edge is $D\phi = 0$, where D is the quasi-material derivative (10.2).

We now study the caudal fin without restriction on angles. Consider a material line coinciding at some time with an arc AB in the plane of the fin edge and normal (Figure 10.3). The arc terminates on either side at the same height but does not cross the fin. If the flow leaves the fin smoothly, an instant later the line is at $A'B'$ immediately downstream of the edge. By conservation of circulation, we have

$$\Gamma' = \Gamma = \int_{AB} \frac{\partial \phi}{\partial s}\, ds = \phi(A) - \phi(B) \equiv \delta\phi = \phi_+ - \phi_- \qquad (10.14)$$

Figure 10.3. At the instant depicted here, the caudal fin sheds streamwise vorticity.

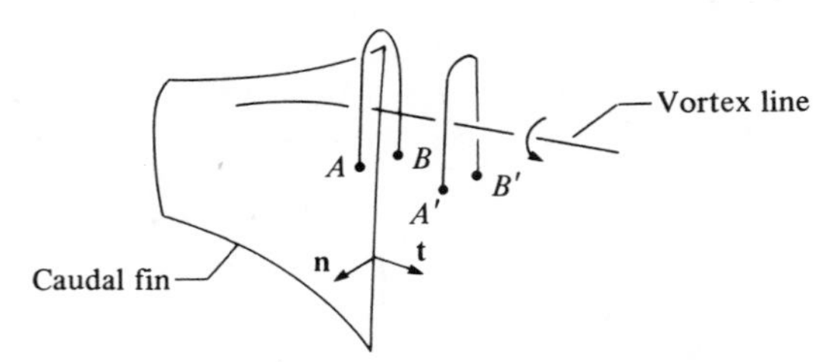

where we now regard the lateral separation of A and B as infinitesimal.

The differentials of Γ at the edge can be studied most conveniently if we assume that there is a neighborhood of the edge where the fin is exactly planar. If vorticity is concentrated there with density γ, we have

$$\gamma = [\mathbf{n} \times \mathbf{u}]_f \tag{10.15}$$

where $[\cdot]_f$ means the jump as the fin is traversed in the direction of $\mathbf{n}$. Owing to the discontinuity, two points A and B adjacent on the plane will separate slightly later. We see, however, that Kelvin's theorem ensures that

$$\frac{d}{dt}(\phi(A) - \phi(B)) - \tfrac{1}{2}(\mathbf{u}_+ - \mathbf{u}_-) \cdot \nabla\phi(A) + \tfrac{1}{2}(\mathbf{u}_+ - \mathbf{u}_-) \cdot \nabla\phi(B) = 0$$

But A and B are material, so that

$$\frac{d\phi}{dt}(A) = \frac{\partial\phi}{\partial t}(A) + \mathbf{u}_+ \cdot \nabla\phi(A)$$

$$\frac{d\phi}{dt}(B) = \frac{\partial\phi}{\partial t}(B) + \mathbf{u}_- \cdot \nabla\phi(B)$$

and therefore

$$\left(\frac{\partial}{\partial t} + \mathbf{u}_m \cdot \nabla\right)\delta\phi = 0, \qquad \mathbf{u}_m = \tfrac{1}{2}(\mathbf{u}_+ + \mathbf{u}_-) \tag{10.16}$$

the last expression holding at the edge in particular.

From (10.15) and (10.16) we then obtain, again at the edge,

$$\gamma = (\mathbf{n} \times \nabla)\delta\phi \tag{10.17a}$$

$$\mathbf{u}_m \times \gamma = -\frac{\partial\delta\phi}{\partial t}\mathbf{n} \tag{10.17b}$$

since $\mathbf{u}_m \cdot \mathbf{n} = 0$. These results hold for any downstream facing edge without restrictions of slenderness or small amplitude. If the last condition is met, however, $\mathbf{u}_m \simeq \mathbf{U}$ and (10.17) gives the explicit vorticity at the edge:

$$\gamma_x = -\frac{\partial\delta\phi}{\partial y}$$

$$\gamma_y = -\frac{1}{U}\frac{\partial\delta\phi}{\partial t}$$

Now $\delta\phi$ is known from the flat-plate solution of Chapter 8 to be

$$2w\sqrt{c^2 - y^2}$$

where $2c$ = depth of fin edge. The energy associated with the shed vorticity is identical to the kinetic energy of the cross section just upstream of the edge, because the vorticity distribution is the same. Thus, energy is shed at the rate $\frac{1}{2}Um(l)w^2$, accounting ultimately for the $-\frac{1}{2}w^2$ term in (10.9).

10.3 Finite-amplitude theory

We retain the slenderness condition and assume as before that vorticity is shed only at a vertical caudal fin edge. The wake S_w is then a material surface of vortex lines, represented in Figure 10.4. (We do not show that S_w is, in fact, highly convoluted well downstream of the tail.)

The present extension to finite amplitudes was accomplished by Lighthill (1975, chap. 5) through an argument based entirely on momentum balance. One reason for giving up energy arguments is that it is no longer obvious how to divide up the rates of working into thrust and lateral components, nor does the velocity **U** have any physical significance. Indeed the "swimming velocity" is an artifice even in the small-amplitude thrust calculation, essentially providing a virtual translation from which work can be deduced.

Let the arc length s be measured along the curved surface of symmetry from the nose to the caudal fin edge ($s = l$), and let $x(s, t)$, $z(s, t)$ denote the intersection of this surface with the xz plane relative to the stationary observer (Figure 10.5). The tangent and normal vectors are then (x_s, z_s) and $(-z_s, x_s)$, respectively, and we let $u = \mathbf{u} \cdot \mathbf{t}$, $w = \mathbf{u} \cdot \mathbf{n}$, where $\mathbf{u} = (\dot{x}, \dot{z})$ is the velocity of a given section of the fish in the xz plane.

Figure 10.4. Notation for the vortical wake.

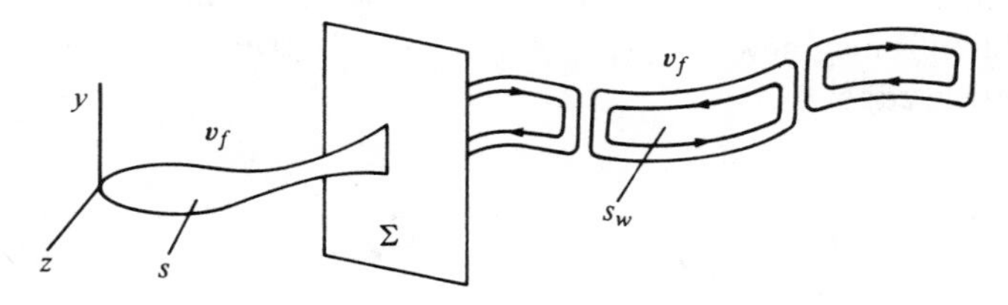

The momentum of a given section is then $mw\mathbf{n}$. The rate of change of total momentum must equal the momentum loss through Σ, minus the pressure force $f_\Sigma \mathbf{t}$ exerted by the fluid on Σ, plus the force F exerted by the fish on the fluid:

$$\frac{d}{dt}\int_0^l mw\mathbf{n}\, ds = [umw\mathbf{n}]_{s=l} - f_\Sigma \mathbf{t} + \mathbf{F} \tag{10.18}$$

Note that (10.18) neglects momentum flux through Σ associated with the small tangential flow through Σ, because in the stationary frame this is negligible for a slender body.

To compute f_Σ, Lighthill uses an argument based directly on Bernoulli's equation. We shall obtain his result by a different method reverting to the energy balance in V_w, which has the advantage of dealing at all times with absolutely convergent integrals, some of which are negligible by slenderness. Let E_w be the wake energy (finite for motion started from rest, say) and consider the stationary observer's energy balance:

$$\dot{E}_w = W_{sw} + W_\Sigma \tag{10.19}$$

where W_{sw} and W_Σ are rates of working on the surfaces S_w and Σ, respectively. Now it is clear that W_{sw} must vanish, because pressure is continuous across the wake.

The rate of change of kinetic energy in V_w is [using (1.11)]

$$\dot{E}_w = \rho \int \nabla\phi \cdot \nabla\dot{\phi}\, dV_w - \frac{\rho}{2}\int (\nabla\phi)^2 u_\Sigma\, d\Sigma$$

where u_Σ is the instantaneous velocity normal to Σ. Using Bernoulli's equation, this may be put into the form

$$\dot{E}_w = \rho \int [p + \tfrac{1}{2}(\nabla\phi)^2]\, \mathbf{t} \cdot \nabla\phi\, d\Sigma - \frac{\rho}{2}\int (\nabla\phi)^2 u_\Sigma\, d\Sigma \tag{10.20}$$

Figure 10.5. Notation for finite-amplitude analysis.

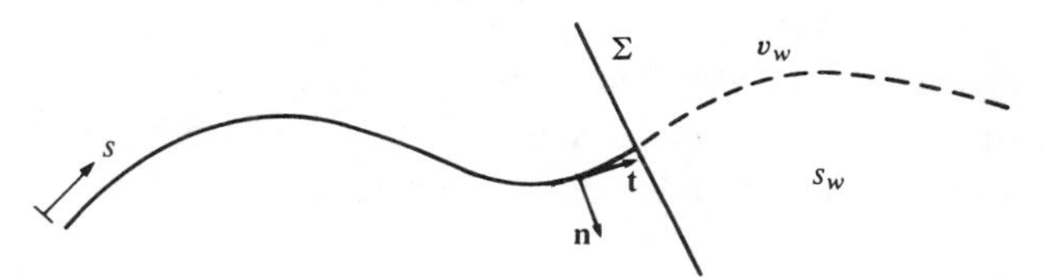

$$\simeq -\frac{\rho}{2}\int (\nabla\phi)^2 u_\Sigma \, d\Sigma \tag{10.21}$$

$$\simeq -\tfrac{1}{2}[mw^2u]_{s=l}$$

Note that (10.20) follows from the smallness of $\mathbf{t}\cdot\nabla\phi$ in slender-body theory, whereas (10.21) comes from the fact that the instantaneous rotation of Σ about the fin edge gives zero contribution, the associated term being an absolutely convergent integral of the product of functions even and odd in the lateral coordinate. This leaves only the effect of parallel translation of the terminal point. From (10.21) we deduce that energy is created in V_w only by vortex shedding in this approximation.

Finally, consider W_Σ in (10.19). This term must be a linear combination of the form

$$W_\Sigma = Au(l, t) + B\omega(l, t) \tag{10.22}$$

where ω is the instantaneous angular speed of Σ. Since (10.21) is independent of ω, we see that $B = 0$ in (10.22), and therefore

$$-W_\Sigma = f_\Sigma u(l, t) = \tfrac{1}{2}[mw^2u]_{s=l}$$

Hence,

$$f_\Sigma = \tfrac{1}{2}[mw^2]_{s=l} \tag{10.23}$$

Returning to (10.18), we now have

$$\mathbf{F} = [-umw\mathbf{n} + \tfrac{1}{2}mw^2\mathbf{t}]_{s=l} + \frac{d}{dt}\int_0^l mw\mathbf{n}\, ds \tag{10.24}$$

Using $-u\mathbf{n} + w\mathbf{t} = \mathbf{j}\times\mathbf{u}$, (10.24) implies Lighthill's equation:

$$(F_x, F_z) = \left[mw(z_t, -x_t) - \frac{m}{2}w^2(x_s, z_s)\right]_{s=l} + \frac{d}{dt}\int_0^l mw(-z_s, x_s)\, ds \tag{10.25}$$

This elegant result, which provides an expression for side force as well as thrust, can be used to study such large-amplitude swimming maneuvers as turning and starting, where the lateral velocity of the fin is momentarily comparable to the swimming velocity [see Weihs (1972, 1973)].

Let us consider a simple lateral progressive wave as an example of large-amplitude motion. If the fish swims to the left with mean speed U, and V is the phase speed of the wave (using notation of chap. 5), then an observer seeing stationary waves is moving with speed $V - U$ relative to the fluid at infinity. As the observer also sees the tangential motion of the fish with velocity $-Q\mathbf{t}$, we see that $w = (U - V)z_s = (U - V)H \cos Qt$ if the coordinate s is measured in units making the wavelength 2π. Then $z_t = -QH \cos Qt$ and $x_s = \alpha$, where $\alpha Q = V$. Therefore,

$$T = \frac{m(l)H^2}{2\alpha}\left[\left(1 - \frac{\alpha^2}{2}\right)V^2 - (\alpha^2 - 1)UV - \tfrac{1}{2}\alpha^2 U^2\right] \quad (10.26)$$

This reduces to a result given earlier in the small-amplitude limit $\alpha \to 1$.

The efficiency may now be defined if the energy shed into the wake, plus UT, is taken to equal the mean rate of working of lateral forces $\langle W_L(t)\rangle$.

$$\eta = \frac{UT}{UT + \Delta W}$$
$$\Delta W = \tfrac{1}{2}\langle[mw^2u]_{s=l}\rangle \quad (10.27)$$

We observe that

$$u(l, t) = (V - U)x_s - Q$$

so that

$$\Delta W = \frac{m(l)}{4\alpha} H^2(V - U)^2[V(1 - \alpha^2) + \alpha^2 U]$$

and if $\theta = U/V$,

$$\eta = \frac{2\theta - \alpha^2(1 - \theta)}{(1 + \theta) - \alpha^2(1 - \theta)} \quad (10.28)$$

Clearly, insofar as efficiency is concerned, the fish can do no better than to make α and θ as close to 1 as possible, reducing the problem to small-amplitude swimming. If the speed is large enough to develop sufficient thrust with small movements, this strategy is optimal. At lower speeds thrust can be maintained by lowering α from 1 (which increases H), thereby significantly increasing the factor in (10.26), and if θ can be maintained close to 1, we see from (10.28) that this thrust is developed without appreciably lowering the efficiency.

10.4 Midbody vortex shedding

The extensions of the foregoing calculations needed to account for the fins and other appendages upstream of the caudal fin require a considerably more elaborate theory, which allows for the presence of free vorticity between the nose and tail. The example we describe below was treated by Lighthill (1975, chap. 4). A related model of laterally thin fish has been developed by Wu (1971*b*) and extended to arbitrary slender forms by Newman and Wu (1973).

We suppose that midbody vortex shedding occurs only at the vertical downstream edge of a slender dorsal fin. The body is again taken to be laterally symmetric.

At first sight it would appear necessary to suppose that lateral excursions of the body be small compared to *depth*, because the vorticity shed by the dorsal fin could eventually lie well away from the surface of symmetry of the fish, thus destroying the lateral symmetry of the problem. In fact, the vortex lines actually remain *in* the surface of symmetry of the fish, within the approximations of slender-body theory. To see this, consider a new lateral coordinate $Z = z - h(x, t)$, in terms of which Bernoulli's equation in the moving frame (with ϕ the perturbation potential) may be written

$$-\frac{p}{\rho} = \phi - h_t\phi_z + \tfrac{1}{2}[(U + \phi_x - h_x\phi_z)^2 + \phi_y^2 + \phi_z^2] - U^2$$

$$= D\phi - h_t\phi_z + \tfrac{1}{2}(\phi_y^2 + \phi_z^2) - h_x\phi_x\phi_z + \phi_x^2$$

In slender-body theory the last two terms are negligible. Dividing $\phi = \phi_e + \phi_o$ into parts even and odd in Z, and noting that $\phi_{ez} = 0$ on the surface of symmetry, the continuity of pressure on the vortex gives

$$D\phi_o + \phi_{oy}\phi_{ey} = 0$$

showing that vortex lines are convected with the flow $(U, \phi_{ey}, 0)$. This is now a property of the plane $Z = 0$, showing that vorticity initially on this plane stays there. (Lighthill refers to the coordinates x, y, z as "stretched straight," because for nonplanar symmetric bodies, the part of ϕ even in Z is the potential for the flow over the fish stretched straight.)

Let the edge of the dorsal fin be at $x = x_f$. Downstream of x_f the potential may be written as $\phi + \tilde{\phi}$, where ϕ satisfies the usual boundary conditions on the body and vanishes on the vortex sheet; $\tilde{\phi}$ has zero normal derivative on the body and takes appropriate values (determined by the Kutta condition given above) on the vortex sheet. As we saw in Chapter 8, the apparent momentum is linear in the potential, and therefore the lateral momentum is $mw + \tilde{m}w_f$, where

$$w_f = w\left(x_f, t - \frac{x - x_f}{U}\right)$$

involves the retarded time. Also, Lighthill notes that since ϕ and $\tilde{\phi}$ are orthogonal in the inner product,

$$(\phi, \tilde{\phi}) = \int \nabla\phi \cdot \nabla\tilde{\phi}\, dV$$

the corresponding kinetic energy is simply $\frac{1}{2}mw^2 + \frac{1}{2}\tilde{m}w_f^2$. The calculations of W_L are therefore readily extended and there results

$$W_L = U\tilde{m}(x_f)w(x_f, t)h_t(x_f, t) + \int_{x_f}^{l} U\tilde{m}'(x)w_f h_t(x, t)\, dx + \cdots \tag{10.5*}$$

$$W_L = \tfrac{1}{2}U\tilde{m}w_f^2\,|_{x=l} + \frac{d}{dt}\int_0^l \tfrac{1}{2}\tilde{m}w_f^2\, dx + \cdots \tag{10.8*}$$

where the terms omitted are those already on the right-hand sides (10.5) and (10.8). It is seen that the dorsal fin generates at x_f a contribution like the bracketed term in (10.9), as would be expected, *plus* the integral term in (10.5*a*). Now, generally $\tilde{m}$ will decrease from some positive value at $x = x_f+$ to zero if and when the upper edge of the fine wake crosses the caudal fin. For closely spaced fins the integral will about cancel the first term on the right of (10.5*); this leaves the energy shed into the wake in (10.8*), but Lighthill suggests that (10.8*) overestimates this loss because mixing of the vortex sheet and the caudal fin boundary layer will occur to some extent. Thus, a row of discrete fins may have little effect on thrust but could produce the lateral momentum of a continuous ribbon fin with smaller wetted area.

Of interest is the case of widely separated dorsal and caudal fins. There the developing phase difference between w_f (convected with speed U) and w (wavelike with phase speed $V > U$, say) can change the sign of the integral in (10.8a), leading to a possibly significant additional source of thrust.

It is probable that detailed analysis of these nonlocal interactions in particular species will shed light on the evolution of fin placement in relation to basic body shape and on numerous special problems such as the propulsion of pectoral fin swimmers (e.g., wrasses).

10.5 Remarks

Breder (1926) divides the main types of swimming of bodies with tails into three groups. The *anguilliform* mode is typified by the eels but, as we have noted elsewhere, this mode characterizes flagellar movements as well. Locomotion may be achieved through resistive or reactive forces, or a combination of the two, the common feature being a well-developed undular movement that propagates down most of the body. The *carangiform* mode (the term comes from the family Carangidae, fishes resembling tunas and mackerels and noted for their speed) is similar, but the movements are more pronounced and restricted to a shorter piece of the tail. The *ostraciiform* mode (from the family Ostraciidae of trunk fishes) is characterized by a small, almost pivoting motion of the rear fin.

In a discussion of the evolution of the carangiform mode, Lighthill (1975, chap. 5) suggests that significant gains in efficiency of reactive swimming can be obtained if the lateral compression that produces the caudal fin is accompanied by a transition to carangiform movements. The argument is based upon the fact that smooth two-dimensional cylindrical shapes will, nevertheless, shed vorticity under prolonged acceleration. In slender-body propulsion, this secondary shedding (which was neglected above) would occur for sufficiently pronounced lateral motion in the anguilliform mode, and thereby feed energy into the wake with little compensating gain in thrust. Such losses are minimized in the carangiform mode, because the extent of tail movement is greatly reduced. On the other hand, as only a fraction of a "wavelength" of the mode may be present at any one time, there is little cancellation of fluctuating side forces in (10.25). The finite-amplitude theory suggests that the sideways recoil from the caudal fin side forces can be minimized by increasing the midbody virtual mass. The needed change in shape is indicated in Figure 10.6*a*.

As the burden of propulsion is shifted to the caudal fin, we can expect an accompanying optimization of its shape. A typical sequence is shown in Figure 10.6*b*. The first stage can be understood from the point of view of slender-body theory, because the scooped-out portion of the fin will be filled with a vortex sheet carrying (ignoring phase differences) the momentum of the tail. Thus, thrust is essentially unaffected, whereas skin friction drag is reduced. In the second stage, slenderness is lost and the fin functions as a sideways flapping "wing." The swimming then acquires characteristics of animal forward flight (see Chapter 11) and can be studied by extending classical methods of unsteady wing theory to the thin, tapered, slightly swept planform typical of many carangiform swimmers. The three-dimensional, small-amplitude theory suggests that the sideways recoil from the caudal fin (1974); he has also studied a complementary large-amplitude two-dimensional problem (Chopra, 1975). We study a related problem in Chapter 11.

We conclude this chapter with a few remarks concerning the viscous resistance of swimming fish. Surprisingly, this resistance can be

Figure 10.6. Transition to the carangiform mode. (*a*) Body shape. (*b*) Caudal fin.

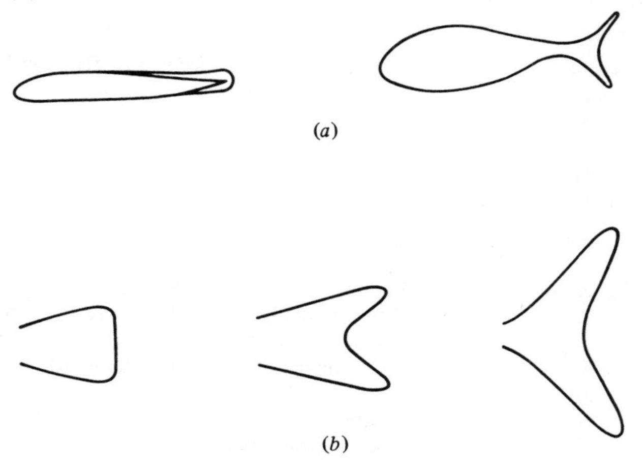

much larger than that of the stretched-straight fish, by a factor of three or more in examples discussed by Bone (1975). This has biological significance when the energetics of swimming is considered. As Weihs (1974) has pointed out, under certain conditions there are significant gains to be had from "burst swimming," the latter involving rapid acceleration in a high-drag swimming mode followed by coasting in a low-drag position.

Fish apparently deal with the matter of resistance with a variety of drag-reducing devices (Bone, 1975). Some of these are:

1. *Surface roughness.* By inducing transition to turbulent flow in the boundary layer, a sizable drag reduction is sometimes possible. Some fish (the larger lunate-tail teleosts) possess thickened scales near maximum depth, which could induce transition before separation occurs. A similar role has been suggested for the denticles of the swordfish.

2. *Vortex generators.* This device is used on aircraft to shed streamwise vorticity at the surface, in order to mix the boundary layer with the adjacent high-speed flow, thereby retarding separation. On the fish *Ruvettus,* for example, a number of small (1-mm) projections occur at points where separation would be likely.

3. *Boundary-layer control.* In some species, one observes over certain areas of the tail a number of subdermal canals that alternately take in and eject fluid as the tail flaps. It is possible that by energizing the boundary layer at the right moment, the discharge of water inhibits flow separation.

4. *Slime.* It has been found that slime on the skin of a barracuda can lead to a 60 percent reduction in turbulent skin friction at a concentration of 1.5 percent.

Exercises

10.1 Verify that (10.10) may be written

$$\langle T\rangle = \tfrac{1}{2}m(l)\langle h_t^2 - U^2h_x^2\rangle$$

10.2 Show that a slender fish can swim by having its body execute a standing wave in the comoving frame but that its efficiency as defined by (10.13) cannot exceed $\frac{1}{2}$.

10.3 Verify that the x component of (10.25) reduces to (10.9) when amplitude is small.

10.4 Show, using (10.25) and assuming the cancellation of side forces, that a pure ostraciiform swimmer with tail movement given by (a)

will not swim according to slender-body theory. Show also that if, instead, the rear edge of the fin remains parallel to the swimming direction ($x_s = 1$ at $s = l$) as in (b) and the edge moves laterally as $\sin \omega t$, the fish will swim according to slender-body theory.

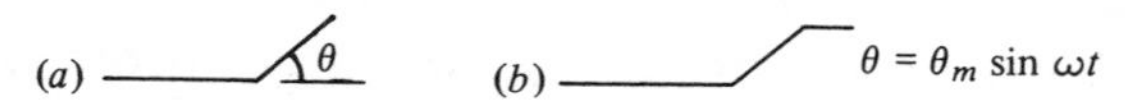

10.5 A well-known result for plane, slender wings at a fixed angle of attack states that sections of the wing aft of the point of maximum span do not contribute to lift. Explain this result assuming that the Kutta condition holds on downstream edges.

10.6 By approximating a continuous curve by a "staircase function" (for example), show that the planar fish shown here will experience a lateral force $-m(x)Dw$ on sections between x_1 and x_2 when the Kutta condition is satisfied on downstream edges of the body.

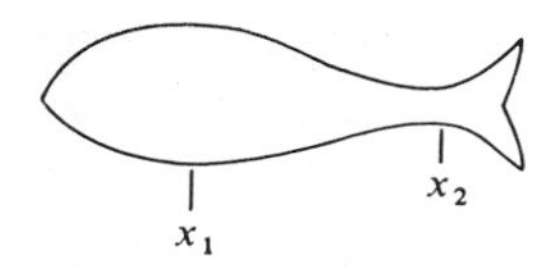

11

Some aspects of the aerodynamics of birds and insects

A principal characteristic of "flight" is that a significant part of the aerodynamic force is needed to cancel the weight of the organism. Thus, certain features of flying apply to buoyant fish. In forward flight such a force can be obtained by creating horizontal bound vorticity, this being the main purpose of the lifting surfaces of the body. The soaring and gliding of birds provides a familiar example where the classical aerodynamics of fixed-wing aircraft can be applied at once, a problem we consider in Section 11.1. It is worth mentioning that there is evidence that observations of birds led Lanchester to the notions of circulation and induced drag of finite wings [see the engrossing historical summary in Durand (1963)].

However, gliding, insofar as it may be taken to be stationary in time, is a rather special instance of animal flight. The common maneuvers of natural flight – takeoff and landing, flapping flight, and hovering – are fundamentally time-dependent phenomena (see the estimate of the frequency parameter σ given in Chapter 1). In effect, natural fliers, in particular the hummingbirds and certain of the hovering insects, explore the problem of lift generation in a very different parameter range from that conventionally exploited in aeronautics. In later sections we examine some of these time-dependent problems in the context of unsteady airfoil theory.

For general references on this subject, see Gray (1968) and the review articles by Lighthill and Weis-Fogh in Wu et al. (1975, vol. 2). For well-illustrated popular accounts of insect flight, see Nachtigall (1968) and Weis-Fogh (1975*b*).

11.1 The finite lifting wing

We treat the finite wing using Prandtl's lifting-line theory [see, e.g., Batchelor (1967, pp. 583–91) and Friedrichs (1966, chaps. 15–18)]. This theory relates the shed vorticity to the lift distribution along the span, leading to a determination of lift and drag in terms of wing geometry. We take the wing to be symmetric with respect to the streamwise x axis (Figure 11.1). The aspect ratio A is defined to be $4b^2/S$, where b is the semispan and S the area of the planform. Roughly, A measures the span in units of a typical chord. Whereas slender-body theory treats wings of low aspect ratio, the problems envisaged here involve a rather large ratio, typically 5 to 15 for many birds. If $A \gg 1$, the bound vorticity of the wing is, on the scale of the span, concentrated near a line called the *lifting line*. However, if circulation around the airfoil is computed at a given station along the span of the wing, one finds a spanwise variation, indicating that the cross-stream vorticity has been shed as streamwise vorticity.

In the limit $A \to \infty$, the problem reduces to that of a lifting-line vortex forming the upstream edge of a vortex sheet. The latter may be assumed to lie in the xy plane. (Because of the self-induced flow field, the sheet is actually deflected downward to some extent; far downstream it is highly convoluted, owing to rolling up at the lateral edges.) Moreover, it is natural to assume that, upon close examination of a given wing section, the local wing loading (lift/unit span) should closely approximate that determined in a two-dimensional airfoil problem involving the local wing profile, local angle of attack, and so on. Van Dyke (1975) has given these two hypotheses of Prandtl's theory a proper mathematical foundation in terms of matched asymptotic expansions in the parameter A^{-1}. The second of these hypotheses has an immediate advantage, in that for a large class of airfoils we have

$$\gamma = kcU\sin(\alpha_{\text{eff}} + \beta) \tag{11.1}$$

where γ is the circulation about the foil, k a geometrical factor, c the chord, α_{eff} the angle of attack, and β the angle of attack at zero lift.

Figure 11.1. Finite lifting wing in the limit $A \to \infty$. (*a*) A finite. (*b*) A infinite.

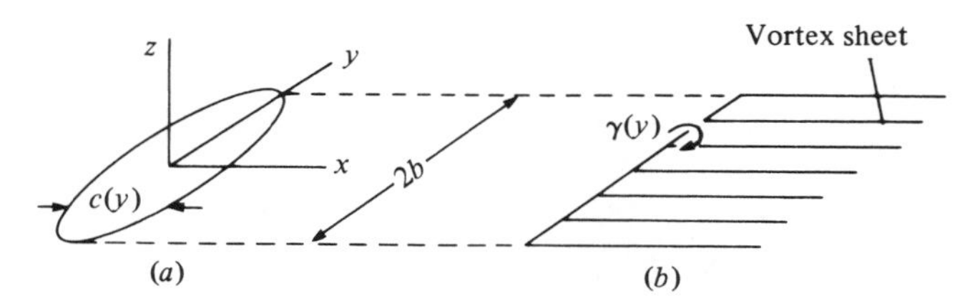

For the finite wing all the parameters may depend upon y, and the notation α_{eff} indicates that the angle of attack has an effective value for any given section.

The value of α_{eff} depends upon the entire vorticity field. In the immediate neighborhood of a point $(0, y, 0)$ on the lifting line, the wing section will have an angle of attack α equal to that determined by the two-dimensional airfoil theory, modified by the flow induced by the vortex sheet. If all angles are small, we then have approximately

$$\alpha_{eff}(y) = \alpha(y) + U^{-1}w^i(0, y, 0) \tag{11.2}$$

where w^i is the vertical velocity component of the induced flow. To compute this quantity at the lifting line, note that minus the stream-wise vorticity shed at section is just

$$\frac{d\gamma}{d\eta} = \gamma'(\eta)$$

Now an infinite vortex tube of unit strength through this section has the velocity field

$$\frac{(0, -z, y-\eta)}{2\pi[x^2 + (y-\eta)^2 + z^2]}$$

which reduces to $[2\pi(y-\eta)]^{-1}$ at $x = z = 0$. The semiinfinite tube will induce half this value, and since the lifting-line vorticity is already accounted for in the local two-dimensional flow field, we have

$$w^i(y) = -\frac{1}{4\pi}\int_{-b}^{+b} \frac{\gamma'(\eta)}{y-\eta}\, d\eta \tag{11.3}$$

Combining the foregoing equations, we obtain

$$\gamma(y) = kcU\left[\alpha + \beta - (4\pi U)^{-1}\int_{-b}^{+b} \frac{\gamma'(\eta)}{y-\eta}\, d\eta\right] \tag{11.4}$$

with $\gamma(\pm b) = 0$, so that γ must be determined by solving an integral equation.

Numerical procedures are most conveniently applied to a slightly different form of (11.4) obtained by defining $y = b \sin \theta$, $\eta = b \sin \theta'$ and setting

$$\gamma = Ub\sum_{n=1}^{\infty} B_n \sin n\theta \tag{11.5}$$

As γ is even in $\theta - \pi/2$, the even-numbered coefficients B_n will vanish. Then (11.4) becomes

$$\sum_{n=1}^{\infty} B_n \sin n\theta = \frac{kc}{b}\left(\alpha + \beta + \frac{1}{4\pi}\sum_{n=1}^{\infty} nB_n I_n\right)$$

where

$$I_n = \int_0^{\pi} \frac{\cos n\theta'}{\cos\theta - \cos\theta'}\, d\theta'$$

The integral may be evaluated as the residue of an integral over the indented unit circle in the complex plane, and one obtains easily

$$I_n = -\pi \frac{\sin n\theta}{\sin\theta}$$

Thus the equation reduces to

$$\sin\theta \sum_{n=1}^{\infty} B_n \sin n\theta = \frac{kc}{b}\left[(\alpha + \beta)\sin\theta - \tfrac{1}{4}\sum_{n=1}^{\infty} nB_n \sin n\theta\right] \qquad (11.6)$$

Given the θ dependence of the various parameters, the coefficients can be computed by approximating both sides as truncated series in sin $n\theta$.

11.1.1 *Lift and drag*

The total lift of the wing is given, according to the Kutta–Joukowski theorem applied to each section, by

$$L = \int_{-b}^{+b} \rho U\gamma\, dy = \frac{\pi}{2}\rho U^2 b^2 B_1 \qquad (11.7)$$

using (11.5). The induced drag D_i may be computed by integrating the streamwise vortex force generated along the lifting line. Thus,

$$D_i = -\rho \int_{-b}^{+b} w\gamma\, dy = \frac{\pi}{8}\rho U^2 b^2 \sum_{n=1}^{\infty} nB_n^2 \qquad (11.8)$$

Note that if among wings of fixed span moving at speed U, one seeks the planform of minimum induced drag capable of developing a given fixed lift (i.e., fixed B_1), then from (11.8) the problem is solved by making all B_n other than B_1 zero. In this case (11.6) reduces to

$$B_1 \sin \theta = \frac{kc}{b}(\alpha + \beta - \tfrac{1}{4}B_1)$$

which would be consistent if, for example, α, β, and k were constant and the planform were *elliptical.*

The lift and drag are usually expressed in terms of coefficients, as follows:

$$L = \tfrac{1}{2}\rho U^2 S C_L$$

where C_L is the lift coefficient, a function of geometry and angle of attack, and

$$D_i = \tfrac{1}{2}\rho U^2 S C_{D_i}$$

where C_{D_i} is the induced drag coefficient. Frequently, one defines a coefficient K by

$$C_{D_i} = \frac{KC_L^2}{A} \tag{11.9}$$

which, with $K = 1/\pi$, is the form taken by C_{D_i} for an elliptical form. The last expression has the advantage of exhibiting the main effect of planform in the factor $1/A$. The value of K depends as well on planform, but only weakly, generally lying within 10 percent of the value for the elliptical wing.

11.1.2 *Gliding at fixed lift*

A gliding bird must adjust its lift and drag to the equilibrium conditions $L = W\cos\theta$ and $D = W\sin\theta$, where W is the weight and θ the glide angle to the horizontal. Thus, glide angle and lift are fixed simultaneously, and the drag must then be adjusted to satisfy the second of the foregoing conditions. In steady soaring, altitude will be maintained when $U\sin\theta$ equals the updraft speed.

Now the total drag is the sum of the drag due to skin friction and that due to vortex shedding. Introducing a skin-friction coefficient C_{D_f} and using (11.9), we have

$$D = \tfrac{1}{2}\rho U^2 S C_{D_f} + \frac{KL^2}{\tfrac{1}{2}\rho U^2 SA}$$

This expression has a minimum speed at the speed U^*,

$$U^* = \left[\frac{4KL^2}{\rho^2 C_{D_f} b^2 S} \right]^{1/4} \tag{11.10}$$

which, as Lighthill (1975, chap. 8) has noted, has some interesting implications, given a bird's ability to alter its planform. Indeed, for *stable* gliding we must assume that $U > U^*$, for only then does an increase in speed lead to a restoring increase of drag. Given that all quantities in (11.11) other than b^2S are essentially constant, U^* can be adjusted by flexing and outstretching the wings (which we may suppose to decrease and increase b and S simultaneously). Some observations discussed by Gray (1968) suggest stable gliding of the black buzzard by the foregoing criterion, and indicate that the tendency for gliding birds to flex their wings when moving into a headwind may be explained as a means of reducing skin-friction drag during rapid flight, where stable gliding may persist even though b^2S is reduced substantially. Finally, to glide stably at *low* speeds, wings should be outstretched, and indeed casual observation of soaring birds and of maneuvers made during approach to landing would suggest that this is the case.

11.2 Unsteady airfoil theory

We consider the unsteady flow about a time-dependent, thin airfoil in accelerated motion. For simplicity, the airfoil will be assumed to have zero thickness, but with time, it may vary in shape, geometrical angle of attack, and vertical position, by small amounts. Considering the flow field in the comoving frame, the velocity U at infinity will be time-dependent. As the resulting potential flow problem is linear and two-dimensional, its analysis is relatively straightforward. We use these results in Section 11.3 to discuss flapping flight and hovering.

Our primary concern will be the lift and thrust developed by the airfoil. If the Eulerian equations of the perturbation velocity are considered in the comoving frame, the acceleration $\rho\dot{U}$ occurs as an added term in the x-momentum balance. This term can be absorbed into the pressure, and the resulting additional drag due to the "pressure" gradient will, by the Archimedean law, vanish because the area of the airfoil is zero.

It is a matter of common experience that unsteady airfoils can produce lift. This lift is due in part to the vorticity bound to the profile. For example, an inclined flat plate brought impulsively to constant speed will eventually experience a constant lift. By the Kutta--Joukowski theorem, there will then be circulation about the airfoil; and by Kelvin's theorem applied to a contour that, before its motion began, encircled the foil, there is a balancing countercirculation in the fluid. To reflect this situation mathematically, we apply the Kutta condition at the trailing edge of the airfoil at each instant. The free vorticity may then be regarded as "shed" from the trailing edge. In the context of thin-airfoil theory, the vorticity is introduced at the point $(a, 0)$, where $2a$ is the chord, and is thereafter convected along the x axis with speed $U(t)$ (Figure 11.2).

The physical validity of the Kutta condition in nonstationary problems deserves scrutiny. If the period of oscillatory components of the motion is smaller than the response time of the viscous boundary layer, vortex shedding cannot keep pace with the airfoil's movements and the Kutta condition is not satisfied. This effect depends on Reynolds number, but in typical problems of animal flight it is probably reasonable to use the Kutta condition provided that the frequency parameter σ (based on a) does not exceed 1. (Viscosity can also alter the distribution of free vorticity when, owing to high frequencies, the gradient of vorticity in the sheet is high, but this places a far less stringent condition upon σ.)

A last condition concerns the convection of free vorticity. Imposition of the Kutta condition at a unique "trailing" edge implies that U must be of one sign (here positive). Then, in the comoving frame, shed vor-

Figure 11.2. Unsteady airfoil in the comoving frame.

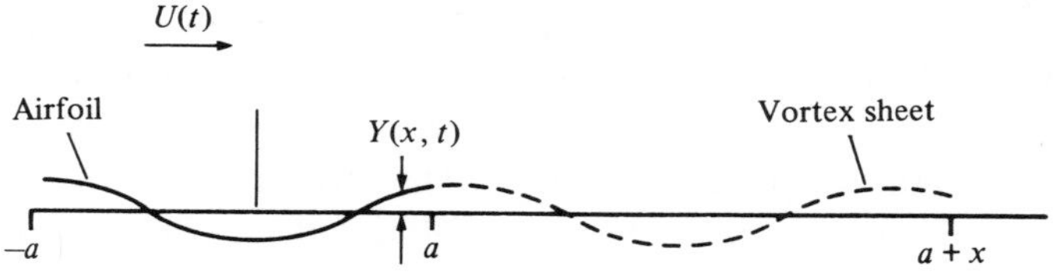

ticity always moves downstream; a vortex shed at time t' will be found at time t a distance $X(t) - X(t')$ downstream of the trailing edge, where

$$X(t) = \int_0^t U(\tau)\, d\tau$$

It will be convenient to divide the perturbation potential into two parts,

$$\phi = \phi_1 + \phi_2$$

where $\phi_1(x, y, t)$ is the potential of the usual instantaneous irrotational flow over the airfoil, determined by the speed $U(t)$ and satisfying the Kutta condition. The part $\phi_2(x, y, t)$ describes the free vorticity *in the presence of the airfoil,* and it too will satisfy the trailing-edge Kutta condition.

Consider first the potential ϕ_2. As ϕ_1 will have already satisfied the inhomogeneous condition ensuring tangential flow at the airfoil, ϕ_2 will satisfy the homogeneous condition

$$\frac{\partial \phi_2}{\partial y} = 0, \qquad |x| < a, \quad y = 0 \tag{11.11}$$

and will jump in value across the vortex sheet

$$[\phi] \equiv \phi_2(x, y+) - \phi_2(x, y-) = c(x, t), \qquad x > a$$

As we take the motion to start from rest at $t = 0$, c will vanish for $x > a + X$, and otherwise will equal the circulation about the segment between x and $a + X$, measured positive for vortices turning the fluid in the counterclockwise sense. It follows that

$$\gamma(\zeta, t) = -\frac{\partial c}{\partial x}(a + \zeta, t)$$

is the density of wake circulation a distance ξ downstream of the trailing edge.

Consider now a single vortex of unit strength at the point $(a + \xi, 0)$. Under the conformal transformation

$$z = \left(\zeta + \frac{a^2}{4\zeta}\right)$$

thus mapping the slit $|x| < a$ onto the circle C of radius $a/2$ at the origin. The point $(a + \xi, 0)$ maps into the point $(b, 0)$, where

$$b = \tfrac{1}{2}(a + \xi + \sqrt{2a\xi + \xi^2})$$

By Cauchy's theorem circulation is invariant under conformal transformation, so that the vortex maps into a unit vortex at $(b, 0)$. We now must add to the vortex flow an image system which, by (11.11), will make C a streamline. To do this, we use the fact that if $f(\zeta)$ is the complex potential of a flow whose singularities lie outside C, then

$$w(\zeta) = f(\zeta) + \bar{f}\left(\frac{a^2}{4\bar{\zeta}}\right)$$

will have C as a streamline (Milne-Thompson, 1955, p. 151). In our case we have

$$w = -\frac{i}{2\pi}\ln(\zeta - b) + \frac{i}{2\pi}\ln\left(\frac{a^2}{4b} - \zeta\right) - \frac{i}{2\pi}\ln\zeta + \frac{i}{2\pi}\ln b$$

so that the image system consists of a countervortex at the image point $(b', 0)$,

$$b' = \tfrac{1}{2}(a + \xi - \sqrt{2a\xi + \xi^2})$$

and a vortex at the origin, the net strength of the image pair being zero.

To satisfy the Kutta condition, we must add to w an additional centered vortex that makes velocity zero at $\zeta = (a/2, 0)$. If the strength of this vortex is κ, we see that

$$\kappa + 1 - (-\xi + \sqrt{2a\xi + \xi^2})^{-1} - (\xi + \sqrt{2a\xi + \xi^2})^{-1} = 0$$

or

$$\kappa = \sqrt{\frac{2a + \xi}{\xi}} - 1$$

and κ also equals the net bound vorticity associated with the isolated downstream vortex. Summing up the effect of wake circulation, we obtain for total bound vorticity associated with ϕ_2,

$$\Gamma_2 = \int_0^X \gamma(\xi, t)\left(\sqrt{\frac{2a + \xi}{\xi}} - 1\right) d\xi \qquad (11.12)$$

Similarly, if Γ_1 is the bound vorticity associated with ϕ_1, then the bound vorticity for the complete flow is

$$\Gamma = \Gamma_1 + \Gamma_2 \tag{11.13}$$

Now, by Kelvin's theorem, Γ must balance net wake vorticity,

$$\Gamma + \int_0^X \gamma(\xi, t)\, d\xi = 0 \tag{11.14}$$

and from (11.12)–(11.14), we obtain

$$\int_0^{X(t)} \gamma(\xi, t) \sqrt{\frac{2a + \xi}{\xi}}\, d\xi = -\Gamma_1(t) \tag{11.15}$$

which is an integral equation for the unknown wake circulation, assuming that $\Gamma_1(t)$ is known [given $U(t)$ and the solution ϕ_1 of the airfoil problem].

As $D\gamma = 0$, we see that γ may be expressed

$$\gamma(\xi, t) = f(X - \xi)$$

and this permits a solution of (11.15) by transform methods. Define

$$G(s) = \int_{t=0}^{t=\infty} e^{-sX(t)}\Gamma_1(t)\, dX(t)$$

$$\hat{Q}(s) = \int_0^\infty e^{-sX} Q(X)\, dX$$

where Q is any variable. Multiplying the left-hand side of (11.15) by e^{-sX}, integrating with respect to X from $t = 0$ to $t = \infty$, then reversing the order of integration gives

$$G(s) = \hat{f}(s) H(s) \tag{11.16}$$

where

$$H(s) = \int_0^\infty e^{-s\xi} \sqrt{\frac{2a + \xi}{\xi}}\, d\xi \tag{11.17}$$

As (11.16) and (11.17) determine $\hat{f}$ in terms of known functions, the solution (γ or c) may be represented as an inverse Laplace transform.

The inversion is particularly simple for impulsive acceleration (translation) to constant speed, for then Γ_1 is constant for $t > 0$ and $G(s) = -\Gamma_1/s$. In this case it can be shown that

$$c(x, t) \sim -\Gamma_1\left(1 - \frac{a}{Ut + a - x}\right), \qquad x \geq a \text{ fixed}, \quad t \to \infty$$

Now $-c(a, t) = \Gamma$ is the circulation about the airfoil, so we see that vorticity is progressively shed and the steady-state circulation is approached after the foil has moved a few chords lengths, the buildup being as in Figure 11.3. However, since the Kutta–Joukowski theorem does *not* hold in unsteady flow, the growth of lift is not the same.

11.2.1 *Lift*

The instantaneous lift will consist of (1) the vortex force $-\rho U\Gamma$, (2) the inertial force determined by the potential flow *without* circulation and without wake vorticity, and, in addition, two new momentum sources different from those discussed in Chapter 8, but similar in that unsteadiness is involved: (3) a contribution L_w from the free vorticity of the wake and (4) a contribution L_c due to unsteadiness of bound vorticity. The last three may be calculated similarly as dP/dt, where

$$P = \rho \int_{-a}^{+a} [\phi]\, dx = m_{12}U + P_w + P_c \tag{11.18}$$

is the downward momentum of the potential flow, $m_{12}(t)$ being an entry in the virtual mass matrix of Chapter 8. To compute P_w, we first consider the contribution of a single wake vortex as was done previously, but with the condition of zero circulation about the airfoil. That

Figure 11.3. Circulation in impulsive acceleration to constant speed U.

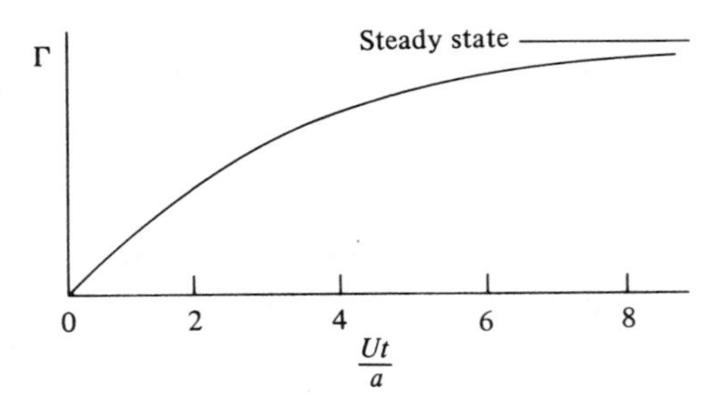

is, in the ζ plane we consider the complex potential w above (11.12). As the circle C is a streamline and w is single-valued outside C, we may write the contribution to P_w in the form (with S denoting the slit)

$$-\oint_s w\,dz = \oint_s z\,dw = 2\,\mathrm{Re}\left[\oint_C \zeta w'(\zeta)\,d\zeta\right]$$

$$= \mathrm{Re}\left[\frac{i}{\pi}\oint_C\left(\frac{\zeta}{\zeta - b'} - \frac{\zeta}{\zeta - b} - 1\right)d\zeta\right]$$

$$= -2b'$$

by the residue theorem. Integrating over the wake, we thus have

$$P_w = -\rho\int_0^X \gamma(\xi, t)(a + \xi - \sqrt{2a\xi + \xi^2})\,d\xi \qquad (11.19)$$

As the complex potential for the circulation is not single-valued, it is simplest to calculate P_c directly as an integral over C:

$$P_c = -\frac{i\rho\Gamma}{2\pi}\int_0^{2\pi}(2i\theta\sin\theta)\left(\frac{a}{2}\right)d\theta = -a\rho\Gamma \qquad (11.20)$$

Gathering these results, (11.14), (11.19), and (11.20) yield the lift

$$l = -\rho U\Gamma + \frac{d}{dt}(m_{12}U) - \frac{d}{dt}\rho\int_0^X \gamma(\xi, t)\,(\xi - \sqrt{2a\xi + \xi^2})\,d\xi$$

$$= \rho U\int_0^X \gamma(\xi, t)\frac{a + \xi}{\sqrt{2a\xi + \xi^2}}\,d\xi + \frac{d}{dt}(m_{12}U) \qquad (11.21)$$

after some reduction using $\partial\gamma/\partial t = -U(\partial\gamma/\partial\xi)$, $\xi > 0$.

The behavior of lift for small time (small X) may be studied by replacing the kernel in (11.15) by $\sqrt{2a/\xi}$ and the right-hand side by $-\Gamma_1(0+)$. The resulting Abel integral equation has the solution

$$\gamma = \frac{\Gamma_1(0+)}{\pi\sqrt{2a}}\frac{1}{\sqrt{X - \xi}} \qquad (11.22)$$

The rather large initial packet of dispersed free vorticity is sometimes referred to as the "starting vortex." Using (11.22) in (11.21), we see that the lift immediately rises to $-\frac{1}{2}\rho U\Gamma_1(0+)$; that is, an impulsively started airfoil immediately acquires, apart from the impulse due to fluid inertia, one-half its final lift. This is a surprising feature of accelerated flow in view of the fact that $\Gamma(0+)$ is zero. For additional details, see Durand (1963, chap. 5) and Friedrichs (1966, chaps. 9–12).

11.3 Thrust realized in flapping flight

We consider the simplest "flapping" mode for the two-dimensional airfoil: a periodic up-and-down motion with fixed orientation. We may assume that the orientation is such that the mean lift is zero, because by linearity finite mean lift (equal to that of the nonflapping airfoil) is acquired by increasing the angle of attack.

We may further assume that $\Gamma_1 = Ae^{i\omega t}$ in (11.13), where A is real, and that

$$\gamma(\xi, t) = \frac{B}{a} e^{i\omega(t-\xi/U)}$$

As

$$\Gamma_t = U \frac{\partial}{\partial a} c(a, t) = -U\gamma(0, t) = -\frac{UB}{a} e^{i\omega t}$$

(11.12) and (11.13) give, with $\sigma = a\omega/U$ and $X \to \infty$,

$$-B = i\sigma(A + BC) \qquad (11.23)$$

where

$$C = F - iG = \int_0^\infty e^{i\sigma u} \left(\sqrt{\frac{2+u}{u}} - 1 \right) du$$

We may solve (11.23) for B:

$$B = \frac{i\sigma A}{1 + i\sigma C} \qquad (11.24)$$

Suppose now that the airfoil is a horizontal flat plate. Adding and subtracting a term from (11.21) and using (11.15), we obtain a suitable expression for the lift (involving integrals convergent as $X \to \infty$). As $m_{12} = 0$, there results

$$l = -\rho U \Gamma_1 - \rho U a \int_0^\infty \frac{1}{\sqrt{2a\xi + \xi^2}} d\xi$$

$$= -\rho U \operatorname{Re} [(A + BD)e^{i\omega t}]$$

where

$$D = P - iQ = \int_0^\infty \frac{e^{i\sigma u}}{\sqrt{u + u^2}}\, du$$

If $v(t) = V \cos \omega t$ is the vertical velocity of the airfoil, then by equation (9.12) (with a replaced by $a/2$) we have $A = 2\pi aV$, so that the mean work done by the lift is

$$\begin{aligned} \langle W_L \rangle &= -\langle lv \rangle = \rho U V^2 \pi a + \rho \frac{UV}{2} \operatorname{Re}(BD) \\ &= \rho U V^2 \pi a [1 - \sigma(pP + qQ)] \end{aligned} \tag{11.25}$$

where we define

$$p = \frac{\sigma F}{(\sigma F)^2 + (1 + \sigma G)^2}$$

$$q = \frac{1 + \sigma G}{(\sigma F)^2 + (1 + \sigma G)^2}$$

When σ is small, we obtain the "quasi-steady" expression $\rho U V^2 \pi a$, which can be explained in terms of the forward tilting of the vortex force in phase with the lateral movement.

The thrust is most easily determined through an energy balance analogous to (9.7),

$$\langle W_L \rangle = UT + \dot{E} \tag{11.26}$$

where $\dot{E}$ is the mean rate of creation of kinetic energy in the wake. To compute $\dot{E}$, note that far downstream of the airfoil, the vortex sheet, idealized as lying on the x axis, has the potential

$$\phi_w = \frac{1}{2\sigma} \operatorname{Re} \left[iBe^{i\omega(t - \xi/U) - (\omega/U)|y|} \right] \operatorname{sgn}(y)$$

The kinetic energy in one wavelength is

$$-\rho \int_0^{2U/\omega} \phi_w \frac{\partial \phi_w}{\partial y}(x, 0+)\, dx = \frac{\rho \pi}{4\sigma^2} |B|^2$$

As $\omega/2\pi$ wavelengths of wake are created per unit time, we have, using (11.24),

$$\dot{E} = \tfrac{1}{2} \rho a^2 \pi^2 V^2 (p^2 + q^2) \tag{11.27}$$

Thus, from (11.25)–(11.27), we obtain

$$T = \rho V^2 \pi a \left[1 - \sigma(pP + qQ) - \frac{\pi}{2}(p^2 + q^2) \right]$$

Using a Bessel function identity, it can be shown (Durand, 1963, p. 307) that T is positive for all σ, although it decreases with increasing σ.

Natural flight involves more complicated wing motions than the above, as well as very involved three-dimensional considerations. Lighthill (1975, chap. 5) considers the two-dimensional problem with

$$Y(x, t) = \alpha \cos \omega t + \beta(x - b) \sin \omega t$$

Here α is the amplitude of up-and-down motion and β is the amplitude of pitching about the point $(b, 0)$. It is found that the larger thrust is realized when $\beta = 0$, at least for moderate σ, but at the price of a lower efficiency than is developed by a combination of the two modes. If b exceeds a by more than a fraction of the chord, a substantial part of the thrust can be traced to leading-edge suction, which part may not be fully realized in practice. A good compromise seems to involve a mixture of comparable modes with the pitching axis forward of the leading edge but rather close to it.

11.4 **Hovering**

Hovering is characterized by little if any horizontal motion of the body. To sustain the weight of the body, surfaces must be moved through the air, and this can be accomplished by moving the wings so that the beat is made in a roughly *horizontal* plane, at an angle of attack that depends upon the strength of the downdraft. The latter flow realizes the downward transfer of momentum taking place at a rate equal to the weight.

Here, as in all other problems of animal flight, the flow field is three-dimensional and very far from the idealizations of linearized theory, but the process can be understood in general terms from quasi-steady finite wing theory, each section of which may be regarded as two-dimensional and quasi-steady. The motions for a wing hinged on a vertical axis are indicated in Figure 11.4*a*. Such wing movement has been termed "normal hovering" by Weis-Fogh (1973) [see also Lighthill (1975, chap. 8)].

The lift computed using these methods is usually close to the weight of the hoverer. However, among certain species of insects, including hoverflies, dragonflies, and small wasps, one finds examples of hovering where the quasi-steady theory is inconsistent in that, to be valid, abnormally high section lift coefficients would have to be realized. In these cases it appears that novel methods of lift generation are at work, involving unsteady aerodynamics.

11.4.1 *The clap and fling*

In attempting to resolve this difficulty in the analysis of the hovering of the small wasp *Encarsia formosa,* Weis-Fogh (1973, 1975*a*) elucidated one such mechanism, termed the "clap and fling." We indicate its main features in Figure 11.4*b*. *Two* wings beat symmetrically about a common hinge axis. The "clap" initiates the process, the wings being momentarily brought together (behind the insect; the body axis is held roughly vertical). The "fling" can be idealized as an opening up of the two wings as if hinged along the lower edge. Bound vorticity, having the sense shown, is generated in each wing and is then carried away as the wings separate and move horizontally.

This remarkable sequence involves two unusual features. In the first

Figure 11.4. Mechanisms of hovering. (*a*) Normal (only one wing shown). (*b*) Clap and fling.

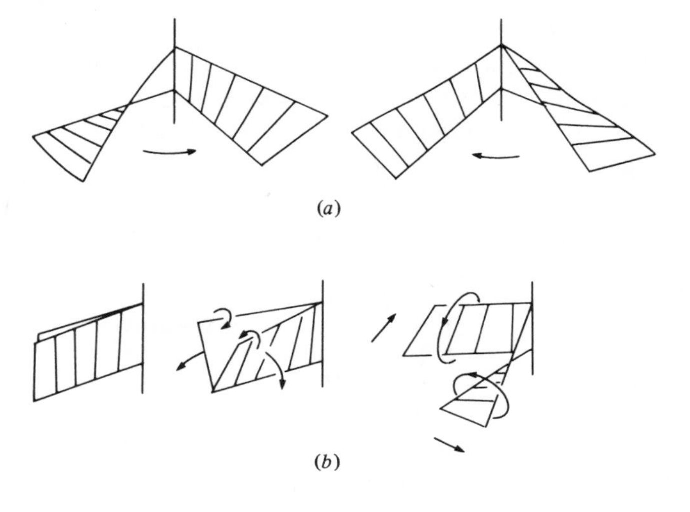

place, Kelvin's theorem is satisfied without, at the moment of separation, the release of any vorticity into the fluid, yet circulation is established. In the second place, circulation is realized independently of lift. If a circular cylinder is placed at the center of a vortex flow, all boundary conditions are satisfied while the cylinder acquires circulation. If the boundary is conformally mapped into any other curve, we generate a class of such circulatory flows. From the point of view of the unsteady theory of Section 11.2, the presence of this "initial" circulation leads to *arbitrary* initial lift, the final lift (in rectilinear motion) being approached *from above.*

Indeed, for a single wing endowed with "initial" circulation, the potential must quickly revert to that normally established once the motion begins and the Kutta condition is established. It is unlikely that this process can be taken to be instantaneous, even though the motion is abrupt, because for the species in question the Reynolds number of the wing is only about 50. From the discussion given above we may deduce that the initial lift then lies about half way between the lift of the initial circulation and the final lift of the airfoil in steady motion.

However, the situation is different with two wings, the problem then being equivalent in inviscid theory to acceleration away from a vertical plane wall. If $t = 0$ marks the end of the fling and the beginning of horizontal motion, then at $t = 0+$ (assuming that the Kutta condition *is* satisfied instantaneously), the "starting vortices" exactly cancel. Thus, the initial lift is that prescribed by the fling. As separation continues, a symmetric field of shed vorticity is generated and (assuming that the wings separate steadily) they eventually act as independent surfaces, thereby realizing the usual lift. For discussion of the theory, see Lighthill (1975, chap. 9) and Ellington (1975).

Exercises

11.1 Verify that for an elliptical planform the coefficient K in (11.9) is $1/\pi$.

11.2 Prove from (11.21) with $m_{12} = 0$ that whenever $\gamma(\xi, t)$, $\xi > 0$, vortex lift never exceeds $-\rho U \Gamma_1(t)$.

11.3 Can an airfoil accelerate from rest in such a way that

$$\max_{t>0} \left[l - \frac{d(m_{12}U)}{dt} \right] > \max_{t>0} |U\rho\Gamma_1|?$$

12

Interactions

In many instances natural swimming and flying involves organisms in sufficiently close proximity to cause significant interaction via the fluid medium. In these cases, the problem emphasized in the preceding chapters – propulsion of a body through a medium at rest – is modified to the extent that a given organism actually moves through the flow field created by its neighbors. To study such interactions, some hypothesis must be made concerning the nature of the response to the modified flow field. A *passive* organism will not modify movements or body geometry in response to environment, even though a modified flow field is sensed. Probably the only examples of a completely passive response are to be found among the microorganisms, and we consider two possible instances below. But the most interesting aspect of this class of problems resides in the capacity of the animal to modify its swimming or flying configuration as it senses changes in the ambient flow field, presumably in order to optimize performance. As we shall see later in the chapter, it has been suggested that such considerations can account for the arrangement of fish in schools and for the formation flying of birds. The observed stability of such groups could be a result of the active orientation of individuals relative to neighbors, but it is more natural to expect that an inherent stability comes from the optimal movements sought by each individual, leading to optimal performance of the group as a whole as members fall into place in a preferred pattern.

The same possibilities apply to a single organism placed in an environment that varies in space and/or time. Examples are thought to be the wave riding of dolphins and certain aspects of the soaring flight of birds. Assuming that the same stability and control mechanisms are available to the single animal as are utilized in interactive locomotion of the type already mentioned, there is every reason to expect that, when possible, use will be made of variations in the ambient flow to increase thrust or otherwise to improve performance.

Finally, it should be noted that interactions between the various fins of a fish, and between the wings of an insect, involve similar considerations, as do the local coordination of cilia movements within a range small compared to the wavelength of the metachronal waves.

12.1 Interaction of waving sheets

Taylor (1951) has commented on Gray's observation that dense clusters of spermatozoans tend to synchronize their movements and to wave their flagella in unison. Taylor went on to suggest that one plausible mechanism for this behavior would be a phase locking caused by interactive stresses. Assuming that the swimmers execute fixed movements, this would then be an example of passive interaction in the Stokesian realm.

This idea was tested by considering a two-dimensional model consisting of two swimming sheets, each movement being as described by equation (3.1), but now with $a_2 = a_1 = 0$ and $\phi_2 = -\phi_1$, where the subscripts refer to the two sheets. If $2h$ is the mean distance between the sheets, the parameter kh now enters into the solution for the flow field, and the average rate of working of one sheet, against stresses on its interior side, can be computed as a function of kh and ϕ_1, $0 \leq \phi_1 \leq \pi/2$. It is found that the required power is smallest when $\phi_1 = 0$ and is largest when $\phi_1 = \pi/2$; the ratio of minimum to maximum varies from 0 to 1 as kh varies from zero to infinity. To examine the direct stability of the interaction, Taylor calculates the correlation between the vertical displacement of the sheet and the internal pressure at the boundary. (For the case $a = 0$, the normal stress is dominated by the contribution from pressure.) It turns out that when the correlation is negative for both sheets, the lifting of the upper sheet is positively correlated with a reduced internal pressure. It is likely that this situation will ultimately reduce the vertical excursion (for fixed available power) and increase the frequency. In this way the phase speed of the wave is increased by the interaction. Similarly, on the lower sheet the frequency will be decreased together with the wave speed, so that the crests of the two waves should drift into phase.

It is reasonable to expect a similar interaction from flagella beating in a common plane, although it is not obvious that such a configuration would be stable in Taylor's sense, because of the additional degree of freedom.

12.2 Bioconvection

We consider next another example where oriented swimming of microorganisms leads to organization of their relative positions, but now the pattern will be on a spatial scale large compared to the size of the organism. The oriented movement will be assumed to be due to a *taxis* (see Chapter 4). We describe a model due to Childress et al. (1975) for pattern formation in a suspension of negatively geotactic organisms. A horizontal layer of such a culture will not remain homogeneous and usually develops an observable pattern consisting of dense clusters of swimmers (see Figure 12.1). In these instances it is found

Figure 12.1. Pattern formation in a 3 mm layer of the dinoflagellate *C. cohnii* at a mean concentration of 10^6 organisms/cm^3. Times: (a) 2 s, (b) 13 s, (c) 19 s, (d) 23 s, (e) 28 s, (f) 32 s, (g) 36 s, (h) 43 s, (i) 2 min. The suspension was swirled initially to render it homogeneous and the early patterns are forming during the decay of the swirl. (Reproduced with permission of the *Journal of Fluid Mechanics.*)

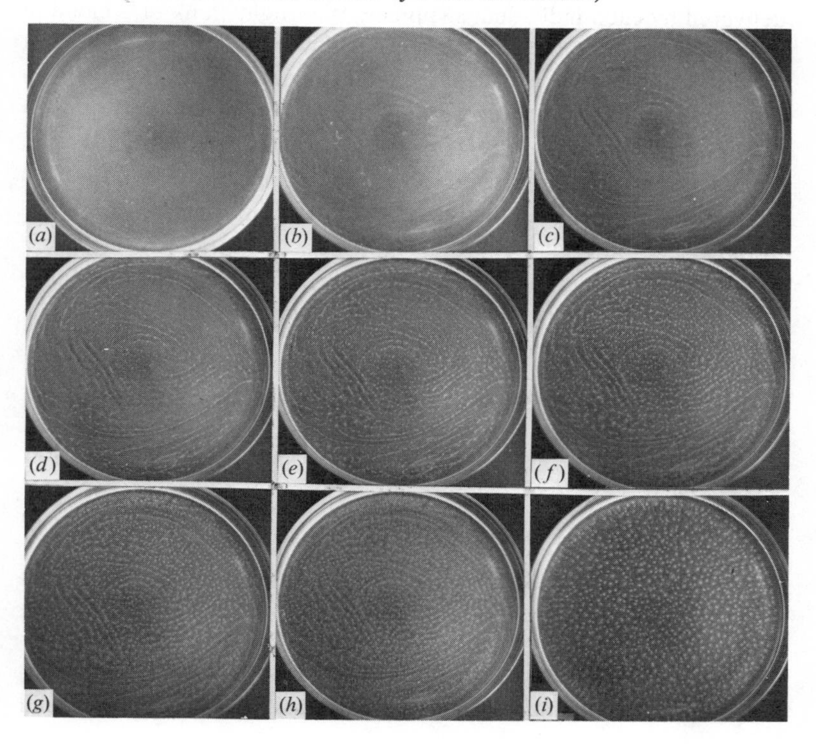

that the mean density of the organisms slightly exceeds that of the medium, so that the upward mean component of their motion (due to the negative geotaxy) tends to cause an accumulation of the relatively heavy biological component near the top of the layer. But this stratification can be unstable when the potential energy released by the downward movement of a parcel of culture is sufficient to overcome the associated viscous dissipation. Such an instability is therefore the viscous counterpart to the classical Rayleigh–Taylor instability of a stratified perfect fluid (as noted by Plesset and Winet, 1974).

It is thought that the observed patterns in cultures of negatively geotactic organisms are largely a consequence of such an instability, although the ultimate fate of an unstable layer will not be obvious from a linear stability calculation. If the negative geotaxy is made the basis for the prolonged organization of the pattern as well as for the onset of instability, then the underlying physical mechanism for organization of the pattern must be the net downward flux of momentum delivered by each individual swimmer. We might think of a cloud of miniature helicopters, each delivering downward momentum at a fixed rate. Placed in a layer of fluid bounded above and below, each rotor sets up a recirculating current, and in suspension a given rotor rides on the combined currents of many neighbors. In particular, it is possible for a rotor to move downward for a time (relative to the horizontal boundaries) even though it continues to move upward relative to the local fluid velocity.

The model we outline now attempts to explain both the instability and the observed patterns in terms of a two-phase continuum. The governing equations are taken to have the form

$$\frac{d\mathbf{u}}{dt} + \frac{1}{\rho}\nabla P - \nu\nabla^2\mathbf{u} = -\mathbf{g}(1 + \alpha c), \qquad \mathbf{g} = g\mathbf{k} \tag{12.1}$$

$$\nabla \cdot \mathbf{u} = 0 \tag{12.2}$$

$$\frac{dc}{dt} + \nabla \cdot \mathbf{J} = 0 \tag{12.3}$$

$$\mathbf{J} = cU(\mathrm{c}, z)\mathbf{k} - \mathbf{D} \cdot \nabla c \tag{12.4}$$

$$D = \kappa_H(\mathbf{i} \cdot \mathbf{i} + \mathbf{j} \cdot \mathbf{j}) + \kappa_V(\mathbf{k} \cdot \mathbf{k}) \tag{12.5}$$

where $\mathbf{u} = (u, v, w)$ is the fluid velocity and c the volume concentration of the microorganisms, the latter being assumed everywhere small compared to 1. The parameter α is a measure of the relative heaviness of the organisms, equal to $(\rho_0/\rho) - 1$, where ρ_0 is the mean density of the organisms and ρ the density of the fluid. From (12.4) and (12.5) we see that the motility is represented as a mean drift upward with speed $U(c, z)$ and effective horizontal and vertical diffusion with coefficients $\kappa_H(c, z)$ and $\kappa_V(c, z)$, respectively. The boundary conditions are that the normal fluid velocity and the flux of organisms through the boundaries both vanish,

$$w = J_3 = 0, \qquad z = 0, \quad -H \tag{12.6}$$

where H is the layer thickness, and that the dynamical constraints [Chandrasekhar (1961) discusses the derivation of these conditions]

$$\frac{\partial^2 w}{\partial z^2} = 0 \qquad \text{on a free boundary} \tag{12.7a}$$

$$\frac{\partial w}{\partial z} = 0 \qquad \text{on a rigid boundary} \tag{12.7b}$$

where by "free" we mean that tangential stress is negligible. Equations (12.1)–(12.7) are similar to the usual Boussinesq model of thermal convection (Chandrasekhar, 1961), the new feature being the self-propulsion of the biological phase.

If we look for time-independent solutions of (12.1)–(12.7) that are independent of the horizontal coordinates x, y, we are led to the integration of

$$cU(c, z) - \kappa_V(c, z)\,\frac{dc}{dz} = 0$$

to obtain an equilibrium concentration profile $c = K(z)$ once the functions $U(c, z)$ and $\kappa_V(c, z)$ are prescribed. For any reasonable choice of these functions, the function $K(z)$ has the expected property of being largest near the top of the layer; if κ_0 and U_0 are typical values of κ_V and U, the organisms tend to be concentrated in a sublayer of thickness $\kappa_0/U_0 = h$, and the dimensionless parameter $\lambda = H/h$ is a useful measure of the relative depth of the layer.

Analysis of the linearization of (12.1)–(12.7) about the sublayer profile, for a number of choices of U, κ_H, and κ_V, shows that, all other parameters being constant (including the mean volume density of the organisms in the homogeneous culture), the sublayer becomes unstable for sufficiently deep layers (i.e., for sufficiently large λ). If the equilibrium is perturbed according to

$$c = K(z) + \epsilon e^{\gamma t} \cos(ax)\, C(z) + O(\epsilon^2) \tag{12.8}$$

(and similarly for the other variables), it is found that at the critical depth, the solution that is first excited is a mode with infinite horizontal wavelength ($a = 0$). This property of the bioconvection problem is quite different from the analogous result for classical thermal convection. Indeed, in the usual Bénard problem, the critical wavelength is finite and comparable to the depth of the layer. The difference lies in the boundary conditions, which require (in effect) that c be given at the walls in the thermal case, whereas in the present problem it is the flux of c that must vanish there.

In addition to the peculiar property of having infinite critical wavelength, there is another, perhaps related result established in unpublished studies by Krishnamurti for the case of small λ and recently derived by Childress and Spiegel (1980) for arbitrary λ. This is that the instability is *subcritical* in the usual sense of nonlinear stability theory. That is, if the nonlinear stability is studied near the critical value λ_c of λ, solutions are found that grow with time for $\lambda < \lambda_c$. One is therefore led to the conjecture that a canonical subcritical bifurcation occurs here, as shown in Figure 12.2. Here A is a measure of the amplitude of stationary patterns, and b is a branch point joining an unstable branch to a stable branch. Observed patterns could then be realized as a jump to the upper stable branch once λ exceeds λ_b. As the solutions on the upper branch are well away from linear modes, they might exhibit the finite pattern wavelengths that are observed (see Figure 12.1*b*).

These conclusions are supported by several partial results. We first

Figure 12.2. Proposed bifurcation curve; u and s denote linearly unstable and stable branches.

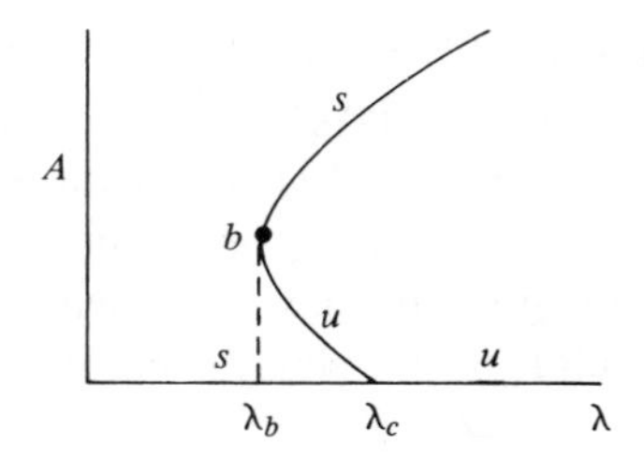

show that when U, κ_H, and κ_V are constants, there is a λ below which the equilibrium sublayer is globally stable in two dimensions. We shall prove that the only solution with finite time and volume mean is the equilibrium profile, provided that λ lies below this critical value. To do this, we use the power-integral approach of Howard (1972). Set $c = K(z) + c'$, where now $K(z) = c_0 e^{z/h}$, $h = \kappa_V/U$, and

$$c_0 = \frac{\lambda c_m}{1 - e^{-\lambda}}$$

c_m being the (fixed) mean organism volume density. The equation for c' is

$$\frac{\partial c'}{\partial t} + \mathbf{U} \cdot \nabla c' + c_0 h^{-1} w e^{z/h} + U\frac{\partial c'}{\partial z} - \kappa_V \frac{\partial^2 c'}{\partial z^2} - \kappa_H \frac{\partial^2 c'}{\partial x^2} = 0 \qquad (12.9)$$

where, for two-dimensional flow, $\mathbf{u} = (\partial\psi/\partial z, 0, -\partial\psi/\partial x)$, ψ being the stream function.

Multiplying (12.9) by c' and integrating over space and time, we obtain, following integration by parts and use of the boundary conditions on c' and ψ,

$$-c_0 h^{-1}[c' w e^{z/h}] + U\left[c'\frac{\partial c'}{\partial z}\right] - \kappa_V\left[\left(\frac{\partial c'}{\partial z}\right)^2\right] - \kappa_H\left[\left(\frac{\partial c'}{\partial x}\right)^2\right] = 0 \qquad (12.10)$$

where $[(\cdot)]$ denotes the time–volume mean,

$$[(\cdot)] = H^{-1}\int_{-H}^{0} \langle\overline{(\cdot)}\rangle \, dz$$

$\overline{(\cdot)}$ being the time mean and $\langle(\cdot)\rangle$ the average over a horizontal plane. Now clearly

$$2U\left[c'\frac{\partial c'}{\partial z}\right] \le \kappa_V\left[\left(\frac{\partial c'}{\partial z}\right)^2\right] + U^2\kappa_V^{-1}[c'^2]$$

and also, by Schwarz's inequality,

$$-c_0 h^{-1}[c' w e^{z/h}] = -c_0 h^{-1}\left[\frac{\partial c'}{\partial x}\psi e^{z/h}\right] \le c_0 h^{-1}\left[\left(\frac{\partial c'}{\partial x}\right)^2\right]^{1/2}[\psi^2]^{1/2}$$

Using these inequalities in (12.10), we have

$$0 \le c_0 h^{-1} \left[\left(\frac{\partial c'}{\partial x} \right)^2 \right]^{1/2} [\psi^2]^{1/2} - \kappa_H \left[\left(\frac{\partial c'}{\partial x} \right)^2 \right] + \tfrac{1}{2} Q \qquad (12.11)$$

where

$$Q = U^2 \kappa_V^{-1} [c'^2] - \kappa_V \left[\left(\frac{\partial c'}{\partial z} \right)^2 \right] \qquad (12.12)$$

But among functions c' satisfying the flux condition in (12.6), Q attains its maximum value, for fixed $[c'^2]$, when c' is a multiple of $e^{z/h}$. This follows easily from a first variation of Q. As the maximum value is zero, we have

$$c_0(\kappa_H h)^{-1} [\psi^2]^{1/2} \ge \left[\left(\frac{\partial c'}{\partial x} \right)^2 \right]^{1/2} \qquad (12.13)$$

Turning now to the momentum equation (12.1), we proceed similarly by multiplying scalarly by **u** and taking the time–volume mean. After some reduction, we obtain

$$\begin{aligned} \Phi &= \nu [(\nabla^2 \psi)^2] \\ &= -g\alpha \left[\psi \frac{\partial c'}{\partial x} \right] \le g\alpha [\psi^2]^{1/2} \left[\left(\frac{\partial c'}{\partial x} \right)^2 \right]^{1/2} \end{aligned} \qquad (12.14)$$

where Φ is the mean viscous dissipation. [Note that (12.14) expresses the fact that the power developed by the swimmers is ultimately converted to heat through viscous dissipation in their wakes.] Now it can be shown that for functions ψ consistent with any permissible combination of boundary conditions, we have

$$[(\nabla^2 \psi)^2] \ge \left(\frac{\pi}{H} \right)^4 [\psi^2] \qquad (12.15)$$

the equality being reached by $\psi = \sin(\pi z / H)$ for the case when both boundaries are free. Using (12.14) and (12.15) in (12.13), we have

$$R \equiv \frac{c_0 H^4 g \alpha U}{\nu \kappa_H \kappa_V} \ge \pi^4 \qquad \text{if} \left[\left(\frac{\partial c'}{\partial x} \right)^2 \right] \ne 0$$

Thus, $[(\partial c'/\partial x)^2]$ necessarily vanishes when $R < \pi^4 = 97.41\ldots$. This amounts to a critical depth condition because, upon introducing c_m,

$$R = \frac{c_m H^4 g\alpha U\lambda}{\nu\kappa_H\kappa_V(1 - e^{-\lambda})}$$

and this is a monotone-increasing function of H. The value 97.41 is not far below the value 120 obtained as the critical value of R for two free boundaries in linear theory (the latter value determining λ_c in Figure 12.2) in the thin-layer limit $\lambda \to 0$. The critical value of R increases with the addition of rigid boundaries, but in these cases the estimate (12.15) can be improved; nevertheless, R_c is already about 313 for $\lambda = 2$ and two free boundaries, so it is apparent that the layer instability is likely to be highly subcritical when $\lambda > 2$. In laboratory experiments the sublayer thickness h is typically 1 mm, and the regularly spaced patterns can be produced in layers with depth greater than 1.5 or 2 mm.

Another approach to the subcritical bifurcation is to seek stationary solutions that might lie on the upper stable branch of Figure 12.2. Childress and Spiegel (1980) have examined this question in a model that neglects the diffusion of the organisms. Again, supposing that U is constant and that the fields are two-dimensional, let us attempt to establish an array of clusters as in Figure 12.1, now as level lines of a scalar function. The streamlines of the fluid motion are the curves ψ = constant, and the trajectories of organisms are the curves $\Psi = \psi - Ux$ = constant. A cluster-type pattern must be a region of closed trajectories (at least in two dimensions), but once organisms are trapped in such a region, it will become a source of downward momentum, producing currents that modify ψ and therefore also Ψ. Thus, the topology of the pattern must be set up in a self-consistent way, and what emerges is a highly nonlinear free boundary problem. In effect, the differential equation for ψ is nonlocal in its dependence on the topology of the solution.†

†Problems of this kind arise in plasma physics, because the plasma trapped within closed lines of force can support currents and hence modify the magnetic field. In this context the term "generalized differential equation" has been used [see, e.g., Grad et al. (1975)].

Approximate solutions have the structure shown in Figure 12.3*a*. The addition of the swimming component to the streamlines of the fluid motion establishes a set of closed orbits where organisms are caught. Observations of the three-dimensional clusters have confirmed that there is a convergence of organisms toward the center at the top of the cluster, and a divergence away from the center at the bottom. Childress and Peyret (1976) have computed such solutions in simulation, using a numerical code able to follow (in two dimensions) 756 particles in a 2 by 1 rectangle. In their calculations, diffusion was incorporated by adding a random walk to the directed motion of geotaxy. An example of a computed pattern is shown in Figure 12.3*b*. Note the formation of a plume of descending organisms near the top of the cluster; when followed farther, these plumes fall through the cluster producing a hole. This phenomenon may explain the small holes that are seen in some clusters (see Figure 3*b* of Childress *et al.*, 1975), and is perhaps not unrelated to plume formation observed among hovering insects (Rigby, 1965).

Figure 12.3. Two-dimensional bioconvection illustrating the formation of a cluster. In (*a*) a typical stream function ψ, when added to the swimming component, produces a pattern of closed trajectories. Sources of downward momentum within this region are then consistent with the assumed ψ. Part (*b*) shows a cluster obtained in the simulation of Childress and Peyret (1976).

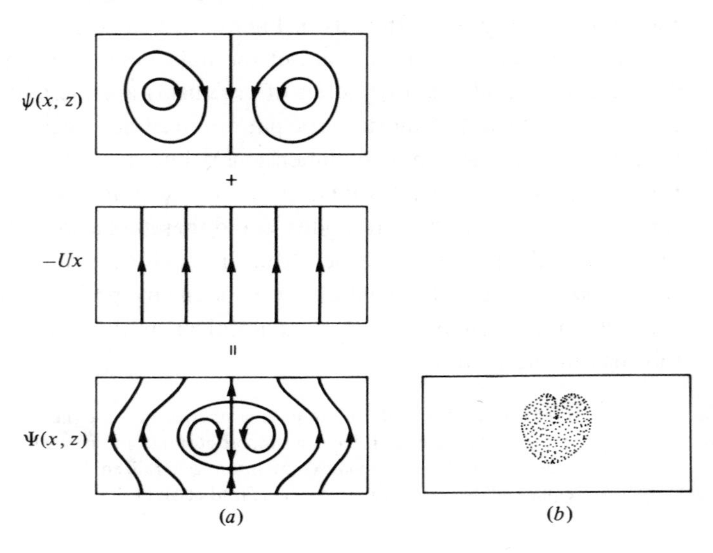

12.3 The schooling of fish

Weihs (1975) has investigated the possible benefits of the interactive swimming of fish in a school by examining the modified environment created by the vortex wakes. If an infinite three-dimensional regular array is considered, the problem can be simplified considerably by imagining that the vortex wakes (see Figure 5.4) may be continued vertically into adjacent identical horizontal layers of fish, so that the wakes may be replaced by vertical sheets of vorticity and the problem becomes two-dimensional. This eliminates certain possible schools where adjacent layers are offset in the horizontal, but avoids the difficulty of following a three-dimensional vorticity field.

We assume next that the two-dimensional vortex sheets roll up, by a nonlinear process observed in other vortex wakes, to produce ultimately an array of two-dimensional point vortices. Assuming that the rolling up is sufficiently rapid, we obtain the point vortex array shown in Figure 12.4, for a single row of fish. Thinking of this row as the first row of a school, we may investigate the optimal position of the fish in the second row.

It is seen that, whether the first-row fish swim in phase or out of it, there is generally a favorable induced flow between two adjacent

Figure 12.4. The vortex wakes studied by Weihs (1975), showing on the left part of a diamond. The parameters b/a and c/a determine a configuration.

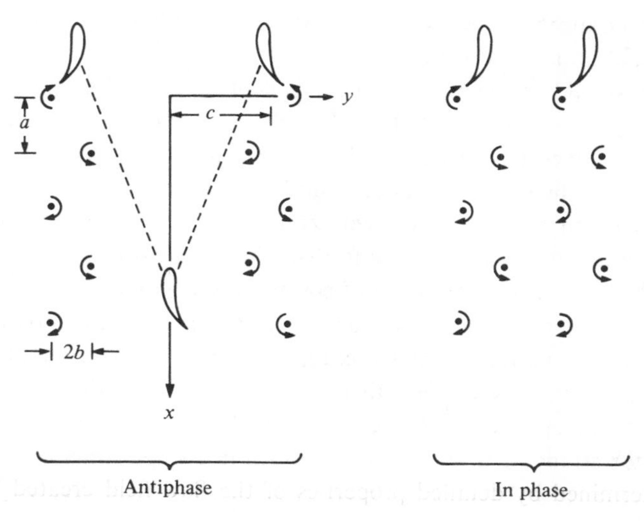

wakes (i.e., a flow in the same direction as the fish is swimming). At the same time the vortices drift downstream with a velocity u_v, which, by the symmetry of the two configurations considered, is a constant. Because we measure this velocity relative to the fluid at infinity, u_v will be positive for a fish producing positive thrust. Along the line $y = 0$ the induced flow is greatest when $x = a \bmod 2a$, and the maximum value is greatest in antiphase swimming. On the other hand, in antiphase swimming the variation of the induced flow with downstream position is greater than for swimming in phase, so that in the former case the second-row fish would presumably have a greater difficulty maintaining position.

There is, however, a still greater penalty to pay when the first-row fish swim in phase, for then there is a net induced lateral flow. Second-row fish must then develop a lateral force in order to maintain position, and accompanying this side force would be a persistent induced drag. It therefore seems likely that the combined advantages of larger induced velocity and lower induced drag in the after rows would cause antiphase swimming to be adopted by the first row.

With this assumption the induced flow is greatest for a streamwise separation of rows of about five times the wavelength of the wake, and Weihs calculates the optimal position of the second row to give a "diamond" (see Figure 12.4) with acute angle 30°. It is suggested that for the school of fish as a whole, the average thrust required to swim at a given speed can be reduced by a factor of one-half or one-third below that of an isolated fish.

It should be remarked that the benefits of schooling vary considerably from row to row. All fish benefit to some extent from the favorable induced flow. Considerable reduction of thrust can be obtained at the second row, but at the third and subsequent odd rows, there may occur partial or complete cancellation of vortices, so that the oncoming flow may be close to that of the first row.

These calculations depend upon some prior knowledge concerning the geometry of the vortex field. Clearly, u_v can be computed given the array and the vortex strength, and thrust can be computed given the same parameters. But thrust must equal drag, which can be estimated independently. Hence, given the array, we may solve for the vortex strength. In practice, however, the configuration of vortices is determined by detailed properties of the flow field created by each fish, and is not uniquely determined by thrust. It would therefore be of interest to study this model in simulation in order to let the vortex field evolve and to investigate the stability of a school under certain constant conditions. Suppose, for example, that all fish are assumed to execute a fixed movement, creating vortices of known strength at fixed times and at points that advance steadily relative to the incident flow. The fish then move relative to one another as they swim until a stable school is established. There is no reason to suppose that swimming movements are invariably independent of the incident flow, but it is not unreasonable to fix the production of vortices as a first approximation.

12.4 Formation flight

Perhaps the most familiar example of collective movement is the formation flying of birds. Lissaman and Shollenberger (1970) have proposed an interesting explanation of the interaction, which indicates that the principal reason for the familiar "vee" formation may be to achieve an equitable distribution of load to the members of the formation while increasing the overall aerodynamic efficiency of the flight. Their model depends upon the possibility that a bird will modify the geometry of its wings so as to optimize the lift distribution, and will maintain the optimal configuration in the presence of interactions with neighbors in a formation. Needless to say, this kind of control is not available to pilots flying fixed-wing aircraft in formation, so in the context of conventional aeronautics, this assumption is unusual and interesting. As we have seen in Chapter 11, the minimum induced drag for a finite fixed wing having prescribed lift is obtained when the lift distribution along the span is elliptical. Noting that the flapping losses should be negligible for the cruising flight of migratory birds, Lissaman and Shollenberger propose that this load distribution is instantaneously maintained and that for the purpose of computing induced drag of the formation, we may disregard the flapping mode.

Actually, the optimal load distribution for an individual within a formation flight is not exactly elliptical. However, calculations establish that the differences are negligible (only a few percent) for all plausible spacings of a row of birds along a horizontal line normal to the direction of flight (line-abreast flight). Now, in line-abreast flight of wings of identical span $2b$ separated by a fixed tip spacing s, the mean reduction in induced drag for a single bird is found to vary from roughly $\frac{1}{2}$ to 1 as $2b/2b + s$ varies from 1 to 0, assuming that the formation consists of three or more birds. The reason for this improvement is easy to understand, because the upwash from neighbors reduces induced drag (by relieving its cause), even as the birds maintain optimal load distributions.

This argument is for line-abreast flight, but a classical theorem of aerodynamics due to Munk (which had an important application to the design of staggered biplanes) establishes that the total induced drag of an array of lifting wings is not altered by displacements of the wings in the direction of flight. However, the *distribution* of induced drag among the various lifting surfaces *is* affected by this staggering. It is found that, whereas in line-abreast flight the middle birds may be able to reduce induced drag by twice the reduction experienced by the tip birds, in a suitable vee formation the distribution can be equalized.

The process of equilibration can be studied by applying lifting-line theory to interacting elliptically loaded wings. The trailing birds, although at a disadvantage in terms of lateral distance from the center of the formation, because of the staggering have gained by "seeing" more of the trailing vortex field created by the other birds. It turns out that the formation which exactly equalizes load is not precisely a vee but rather a curved vee swept back more severely at the tips than at the apex. It is also found that only the nearest six or eight birds significantly affect the lift distribution of an individual, so that the vee need not be symmetrical to be effective.

It is likely that under conditions of fixed power output, the vee formation would be set up quite naturally as new members join the formation and move into position, for a given bird would sense the increased effort that would be required to maintain speed, were it to move forward of a developing line. We must suppose that a bird which attempts to drift within the vee, so as to further reduce drag, would not be tolerated, in which case a stabilizing behavior would be to fly at fixed power. An example of the optimal configuration is shown in Figure 12.5.

12.5 Other examples

The spatial and temporal changes in the local fluid environment can be utilized quite generally as a source of energy for movement. A case in point is the wave riding of dolphins. Hayes (1953)

Figure 12.5. Optimal formation for a flight of nine birds with tip spacing = $\frac{1}{4}$ span, according to Lissaman and Shollenberger (1970).

noted that the mechanism exploited by dolphins probably involves a force balance similar to that shown in Figure 12.6. The drag D and weight W of the fish are balanced by a force $B = P + L$, due to the local pressure gradient and to the lift. The surface of the water is one of constant pressure, and assuming that the fish is near the surface, the pressure gradient force P is approximately normal to the swimming direction and given by

$$P = W_0 \cos \theta + C$$

where W_0 is the mass of displaced water and C is the centrifugal force due to streamline curvature. The important point is that the streamwise force balance gives

$$D = W \sin \theta$$

showing that the principal advantage of wave riding comes from the absence of a component of the pressure force in the direction of swimming. Since the drag is balanced by the streamwise component of the weight vector of the body, rather than by simply the streamwise component of the *excess* weight, the fish is "pulled downhill" as if it were

Figure 12.6. Wave riding.

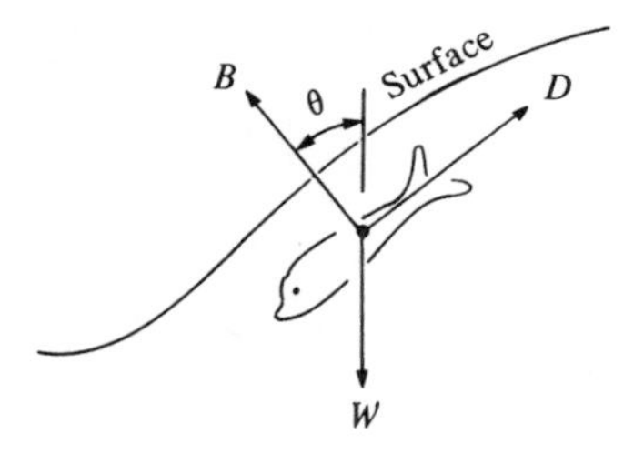

not submerged. The same principle would, of course, apply to surf riding, although there the term $W_0 \cos\theta$ would then be absent. Also, a similar principle might apply to the ridge soaring of birds provided that the pressure gradient has the assumed inclination to the flight path; here, also, the aerodynamic lift dominates the buoyancy force.

A somewhat different class of problems involves the active modification of body geometry to take advantage of random fluctuations in the ambient flow field. Wu (1972) and Hendriks (1975) have discussed this aspect of interactive flying in the case of a lifting two-dimensional surface in a fluctuating oncoming stream. The problem of determining the optimal movement of the wing (minimizing power required for fixed thrust), given the direction and strength of the incident flow as a function of time, is an involved problem of unsteady wing theory, similar in spirit to the calculation of accelerated motion discussed in Chapter 11. It is, of course, possible, and perhaps essential, that the allowed movements of the wing involve changes of wing foil as well as heaving and pitching.

The problem is complicated by the fact that two time scales are involved, one determined by the flapping rate, the other by the time required to move through a gust of typical scale. If both processes are so slow that the wing operates in the quasi-steady mode, then within linear theory the effect of varying angle of attack can be studied independently and the corresponding thrust enhancement can be understood in terms of the tilting of the lift vector.

REFERENCES

Note: SF1 and *SF2* refer to volumes 1 and 2 Wu et al. (1975).

Adler, J., 1975. Chemotaxis in bacteria. Annu. Rev. Biochem. *44*:341–56.

Batchelor, G. K., 1967. *An Introduction to Fluid Dynamics.* Cambridge University Press, New York.

Batchelor, G. K., 1970. Slender-body theory for particles of arbitrary cross-section in Stokes flow. J. Fluid Mech. *44*:419–40.

Berg, H. C., 1975*a*. Bacterial movement. In *SF1,* pp. 1–11.

Berg, H. C., 1975*b*. Chemotaxis in bacteria. Annu. Rev. Biophys. Bioeng. *4*:119–36.

Berg, H. C., 1976. How spirochetes may swim. J. Theor. Biol. *56*:269–73.

Blake, J. R., 1971*a*. Infinite models for ciliary propulsion. J. Fluid Mech. *49*:209–27.

Blake, J. R., 1971*b*. A spherical envelope approach to ciliary propulsion. J. Fluid Mech. *46*:199–208.

Blake, J. R., 1972. A model for the micro-structure in ciliated organisms. J. Fluid Mech. *55*:1–23.

Blake, J. R., 1974. Hydrodynamic calculations on the movements of cilia and flagella, parts I and II. J. Theor. Biol. *45*:183–203; *52*:67–82.

Blake, J. R., and Sleigh, M. A., 1975. Hydromechanical aspects of ciliary propulsion. In *SF1,* pp. 185–209.

Blakemore, R., 1975. Magnetotactic bacteria. Science *190*:377–9.

Bone, Q., 1975. Muscular and energetic aspects of fish swimming. In *SF2,* pp. 493–528.

Breder, C. M., 1926. The locomotion of fishes. Zoologica *4*:159–297.

Brennen, C., 1974. An oscillating boundary-layer theory for ciliary propulsion. J. Fluid. Mech. *65*:799–824.

Brennen, C., 1975. Hydromechanics of propulsion for ciliated microorganisms. In *SF1,* pp. 235–51.

Chandrasekhar, S., 1961. *Hydrodynamic and Hydromagnetic Instability.* Oxford University Press, London.

Cheung, A. T. W., and Winet, H., 1975. Flow velocity profile over a ciliated surface. In *SF1,* pp. 223–4.

Childress, S., and Peyret, R., 1976. A numerical study of two-dimensional convection by motile particles. J. Mec. *15*:753–79.

Childress, S., and Spiegel, E. A., 1980. Pattern formation in a suspension of swimming microorganisms: nonlinear aspects. Submitted to J. Fluid Mech.

Childress, S., Levandowsky, M., and Spiegel, E. A., 1975. Pattern formation in a suspension of swimming microorganisms: equations and stability theory. J. Fluid Mech. *63*:591–613.

Chopra, M. G., 1974. Hydromechanics of lunate-tail swimming propulsion. J. Fluid Mech. *64*:375–91.
Chopra, M. G., 1975. Lunate-tail swimming propulsion, In *SF2,* pp. 635–50.
Chwang, A. T., and Wu, T. Y., 1971. A note on the helical movement of micro-organisms. Proc. R. Soc. Lond. *A178*:327–46.
Cox, R. G., 1970. The motion of long slender bodies in a viscous fluid. Part I. General theory. J. Fluid Mech. *44*:791–810.
Durand, W. F., 1963. *Aerodynamic Theory,* vol. 2. Dover Publications, New York.
Eckert, R., 1972. Bioelectric control of ciliary activity. Science *176*:473–81.
Ellington, C. P., 1975. Non-steady-state aerodynamics of the flight of *Encarsia formosa.* In *SF2,* pp. 783–96.
Freidrichs, K. O., 1966. *Special Topics in Fluid Dynamics.* Gordon and Breach, New York.
Grad, H., Hu, P. N., and Stevens, D. C., 1975. Adiabatic evolution of plasma equilibrium. Proc. Natl. Acad. Sci. U.S.A. *72*:3789–93.
Gray, J., 1968. *Animal Locomotion,* chap. 9. W. W. Norton & Company, New York.
Gray, J., and Hancock, G., 1955. The propulsion of sea-urchin spermatozoa. J. Exp. Biol. *32*:802–14.
Hancock, G., 1953. The self-propulsion of microscopic organisms through liquids. Proc. R. Soc. Lond. *A217*:96–121.
Hauser, D., Levandowsky, M., and Glassgold, J., 1975. Ultrasensitive chemosensory responses by a protozoan to epinephrine and other neurochemicals. Science *190*:285–6.
Hayes, W. D., 1953. Wave riding of dolphins. Nature *172*:1060.
Hendriks, F., 1975. Soaring birds as "Maxwell demons." In *SF2,* pp. 975–84.
Honda, H., and Miyake, A., 1975. Taxis to a conjugation-inducing substance in the ciliate *Blepharisma.* Nature *257*:678–9.
Howard, L. N., 1972. Bounds on flow quantities. Annu. Rev. Fluid Mech. *4*:473–494.
Jahn, T. L., and Votta, J. J., 1972. Locomotion of protozoa. Annu. Rev. Fluid Mech. *4*:93–116.
Keller, J. B., and Rubinow, S. R., 1976. Swimming of flagellated microorganisms. Biophys. J. *16*:151–70.
Keller, S. R., Wu, T. Y., and Brennen, C., 1975. A traction-layer model for ciliary propulsion. In *SF1,* pp. 253–71.
Lamb, H., 1932. *Hydrodynamics,* 6th ed. Cambridge University Press, New York.
Landau, L. D., and Lifschitz, E. M., 1959. *Fluid Mechanics.* Pergamon Press, Elmsford, N.Y.
Lighthill, J. L., 1975. *Mathematical Biofluiddynamics,* SIAM vol. 17. Regional conference series in applied mathematics. Society for Industrial and Applied Mathematics, Philadelphia.
Lighthill, J. L., 1976. Flagellar hydrodynamics. SIAM Rev. *18*:161–230.
Liron, N., and Mochon, S., 1976. The discrete-cilia approach to propulsion of ciliated micro-organisms. J. Fluid Mech. *75*:593–607.
Lissaman, P. B. S., and Shollenberger, C. A., 1970. Formation flight of birds. Science *168*:103–5.

Milne-Thompson, L. M., 1955. *Theoretical Hydrodynamics,* Macmillan, New York.

Nachtigall, W., 1968. *Insects in Flight.* McGraw-Hill Book Company, New York.

Naitoh, Y., 1974. Bioelectric basis of behavior in protozoa. Am. Zool. *14*:883–93.

Newman, J. N., and Wy, T. Y., 1973. A generalized slender-body theory for fish-like forms. J. Fluid Mech. *57*:673–93.

Parducz, B., 1967. Ciliary movement and coordination in ciliates. Int. Rev. Cytol. *21*:91–128.

Plesset, M. S., and Winet, H., 1974. Bioconvection patterns in swimming microorganism cultures as an example of Rayleigh-Taylor instability. Nature *248*:441–3.

Prandtl, L., 1952. *Essentials of Fluid Mechanics.* Hafner Publishing Co., New York.

Purcell, E. M., 1977. Life at low Reynolds number. Am. Jour. Phys. *45*:3–11.

Rigby, M., 1965. Convection plumes and insects. Science *150*:783.

Roberts, A. M., 1970. Geotaxis in motile microorganisms. J. Exp. Biol. *53*:687–99.

Saffman, P. G., 1967. The self-propulsion of a deformable body in a perfect fluid. J. Fluid Mech. *28*:385–9.

Shack, W. J., Fray, C. S., and Lardner, T. J., 1974. Observations on the hydrodynamics and swimming motions of mammalian spermatozoa. Bull. Math. Biol. *36*:555–65.

Sleigh, M. A., ed., 1974. *Cilia and Flagella.* Academic Press, New York.

Sommerfeld, A., 1952. *Lectures on Theoretical Physics,* vol. 2. *Mechanics of Deformable Media,* Academic Press, New York.

Strickler, J. R., 1975. Swimming of planktonic *Cyclops* species (copepoda, crustacea): pattern, movements and their control. In *SF2,* pp. 599–613.

Taylor, G. I., 1951. Analysis of the swimming of microscopic organisms. Proc. R. Soc. Lond. *A209*:447–61.

Taylor, G. I., 1952. The action of waving cylindrical tails in propelling microscopic organisms. Proc. R. Soc. Lond. *A211*:225–39.

Tuck, E. O., 1968. A note on the swimming problem. J. Fluid Mech. *31*:305–8.

Van Dyke, M., 1975. *Perturbation Methods in Fluid Mechanics,* chap. 9. The Parabolic Press, Stanford.

Weihs, D., 1972. A hydrodynamical analysis of fish turning maneuvers. Proc. R. Soc. Lond. *B182*:59–72.

Weihs, D., 1973. The mechanism of rapid starting of slender fish. Bioheology *10*:343–50.

Weihs, D., 1974. Energetic advantages of burst swimming of fish. J. Theor. Biol. *48*:215–29.

Weihs, D., 1975. Some hydrodynamical aspects of fish schooling. In *SF2,* pp. 703–718.

Weis-Fogh, T., 1973. Quick estimates of flight fitness in hovering animals, including novel mechanisms of lift production. J. Exp. Biol. *59*:169–230.

Weis-Fogh, T., 1975*a*. Unusual mechanism for the generation of lift in flying animals. Sci. Am. *233*(5):80–7.

Weis-Fogh, T., 1975*b*. Flapping flight and power in birds and insects, conventional and novel mechanisms. In *SF2,* pp. 729–62.

Winet, H., 1974. Geotaxis in protozoa I. A propulsion-gravity model for *Tetrahymena* (ciliata). J. Theor. Biol. *46*:449–65.

Wolken, J. J., 1971. *Invertebrate Photoreceptors*. Academic Press, New York.

Wu, T. Y.-T., 1971*a*. Hydromechanics of swimming of fishes and cetaceans. Adv. Appl. Mech. *11*:1–63.

Wu, T. Y.-T., 1971*b*. Hydromechanics of swimming propulsion. Part 3. Swimming and optimum movements of slender fish with side fins. J. Fluid Mech. *46*:545–68.

Wu, T. Y.-T., 1972. Extraction of flow energy by a wing oscillating in waves. J. Ship Res. *14*:66–78.

Wu, T. Y.-T., Brokaw, C. J., and Brennen, C., eds., 1975. *Swimming and Flying in Nature,* vols. 1 and 2, Plenum Press, New York.

INDEX